MULTILEVEL NETWORKS IN EUROPEAN FOREIGN POLICY

T0299939

Multilevel Networks in European Foreign Policy

ELKE KRAHMANN

Routledge
Taylor & Francis Group

LONDON AND NEW YORK

First published 2003 by Ashgate Publishing

Reissued 2018 by Routledge
2 Park Square, Milton Park, Abingdon, Oxon OX14 4RN
711 Third Avenue, New York, NY 10017, USA

Routledge is an imprint of the Taylor & Francis Group, an informa business

A Library of Congress record exists under LC control number: 2002102837

ISBN 13: 978-1-138-71608-7 (hbk)
ISBN 13: 978-1-138-71606-3 (pbk)
ISBN 13: 978-1-315-19719-7 (ebk)

Contents

List of Figures   *vi*
List of Tables   *vii*
Preface   *viii*
List of Abbreviations   *x*

1   From National Foreign Policy to Multilevel Networks   1

2   Multilevel Network Theory   17

3   European Union: The Dual-Use Control Agreement   43

4   Transatlantic Community: Air Strikes in Bosnia   79

5   United Kingdom: The Tactical Air-to-Surface Missile   113

6   Conclusion   147

Appendix   *165*
Bibliography   *179*
Index   *187*

List of Figures

3.1	Proportion of preference changes	71
3.2	No change of preferences	72
3.3	Unclear/undecided preferences	73
3.4	Change of preferences	74
4.1	Proportion of preference changes	102
4.2	No change of preferences	102
4.3	Unclear/undecided preferences	103
4.4	Change of preferences	104
4.5	Blocked preference changes	105
5.1	Proportion of preference changes	138
5.2	No change of preferences	138
5.3	Unclear/undecided preferences	139
5.4	Change of preferences	140
5.5	Blocked preference changes	141
6.1	Proportion of preference changes	148
6.2	Proportion of preference changes - curve fit	149
6.3	No change of preferences	150
6.4	Change of preferences	151
6.5	Unclear/undecided preferences	152
6.6	Blocked preference changes	153

List of Tables

2.1	Matrix of dyadic power relations	22
2.2	Rational action	38
3.1	Descriptive statistics	75
4.1	Descriptive statistics	106
5.1	Descriptive statistics	142
6.1	Descriptive statistics	154

Preface

European foreign policy has faced a number of transformations over the past decades. Two countervailing trends have influenced the writing of this book. On the one hand, international organizations such as the European Union, NATO and the OSCE have expanded their role in the making of foreign policy in the region. On the other hand, public and private actors at the national and subnational levels increasingly engage directly in European foreign policy through a network of transnational relations. As a consequence, European foreign policy making has become more and more fragmented - both in terms of actors and levels of analysis.

International relations theory has responded to this challenge with increased multilevel theorizing. In particular, two sets of approaches can be discerned. One focusses on the study of the European Union and its member states. It contends that the foreign policy making process of the European Union is unique because of the close linkages between actors at the national and international levels within an ever increasing framework of institutions. Accordingly, the proponents of this approach argue that European foreign policy analysis needs its own distinct multilevel theories, such as policy network analysis. The other set of approaches, which includes transnationalism and the two-level game, perceive the growing fragmentation of foreign policy making capabilities among public and private actors as a general feature of contemporary foreign policy making in North America and Europe and analyses aspects of it, such as international negotiations or transnational linkages, across states and regions.

In *Multilevel Networks in European Foreign Policy*, I attempt to develop and illustrate the utility of a multilevel theory which seeks to cross both perspectives. Specifically, I argue that European foreign policy cannot be disconnected from its broader international context, such as Europe's relations with the United States and the other international organizations play a crucial role in the definition of foreign policies in Europe. However, I also contend that multilevel theorizing should move beyond the analysis of specific sets of relations, such as transnational linkages or domestic-international influences, towards the integration of the national, transnational and international level within one theoretical framework.

In *Multilevel Networks in European Foreign Policy*, I seek to demonstrate that the British policy network approach is especially suited for the development of such a multilevel theory of European foreign policy. Notably, the approach has already been successfully applied to the national, transnational and international levels in the study of European decision-making. However, few studies combine all three level of analysis. Moreover, the British policy network approach has been criticised for its lack of testable hypotheses which explain decision-making processes. In this book, therefore, I propose a number of modifications in order to improve the explanatory capabilities of the British policy network approach. Specifically, I suggest combining the approach with rational choice hypotheses concerning the way in which network actors might use their national, transnational and international relations to change each others' foreign policy preferences and thus the outcomes of the European foreign policy process.

I then proceed to test the proposed 'multilevel network theory' in three case studies of European foreign policy decision-making within different contexts. The first case study examines European foreign policy making within the European Union. In particular, it analyses the making of common European export controls on goods with civil and military applications ('dual-use'). The second case study investigates the role of the transatlantic community within the European foreign policy decision-making process in the case of air strikes in Bosnia. And the final case study examines the making of national foreign policies in Europe and how they are influenced by the broader European foreign policy network in the example of the British tactical air-to-surface missile project. In the conclusion of this book, I attempt to offer an evaluation of the proposed multilevel network theory and its insights into the multilevel nature of European foreign policy making.

In the writing of this book I have profited from the help of many friends and colleagues to whom I would like to express my gratitude. I am particularly indebted to Prof. William Wallace, Prof. Helga Haftendorn, Prof. Margot Light and Prof. John Peterson who provided valuable comments and criticism. I also would like to thank the Deutsche Forschungsgesellschaft, the German Academic Exchange Service, the British Council and the ESRC who funded my research.

List of Abbreviations

BDI	Federation of German Industries
CDU	Christian Democratic Union
CFSP	Common Foreign and Security Policy
CND	Campaign for Nuclear Disarmament
COCOM	Coordinating Committee for Multilateral Export Controls
CSCE	Conference for Security and Cooperation in Europe
CSU	Christian Social Union
DGB	German Labour Union Association
DIHT	German Chambers of Commerce and Industry
DTI	Department of Trade and Industry
EAPC	Euro-Atlantic Partnership Council
EC	European Community
EDIG	European Defence Industry Group
EEC	European Economic Community
EU	European Union
FDP	Free Democratic Party
INF	Intermediate-Range Nuclear Forces
MoD	Ministry of Defence
MPs	Members of Parliament
NACC	North Atlantic Cooperation Council
NATO	North Atlantic Treaty Organization
NPG	Nuclear Planning Group
OECD	Organisation for Economic Co-operation and Development
OSCE	Organization for Security and Cooperation in Europe
PDS	Party of Democratic Socialism
PfP	Partnership for Peace
RAF	Royal Air Force
SPD	Social Democratic Party
TASM	Tactical Air-to-Surface Missile
UN	United Nations
UNICE	Union of Industrial and Employer's Confederation of Europe
UNPROFOR	United Nations Protection Force
US	United States of America
VDMA	German Association of Machinery and Plant Manufacturers
WEU	Western European Union

1   From National Foreign Policy to Multilevel Networks

Introduction

In the 1990s a growing consensus has emerged in the analysis of foreign policy decision-making according to which it has become necessary to move from single-level approaches in international relations and foreign policy analysis towards a theoretical integration of the domestic, transnational and international levels of analysis (Müller and Risse-Kappen, 1993: 47). The call for multilevel approaches originates from the observation that foreign policy decision-making in Europe and North America has become increasingly integrated since the Second World War. As a consequence, it has been suggested that governments are unable to unilaterally control their foreign or even domestic affairs, as presumed by single-level models of foreign policy decision-making. Conversely, foreign policy making appears to be influenced by a broad variety of public and private actors at the national, transnational and international levels of analysis.

Several analytical frameworks have been proposed which combine different levels of analysis. Specifically in Europe, where the trend towards a fusion of decision-making processes has been recognized in the context of the European Union (EU), multilevel approaches have become increasingly popular. However, many of these models apply only to the specific context of EU institutions. Indeed, some authors have argued that the integrated foreign policy making process among EU member states differs from transnational and international decision-making in other contexts and, therefore, requires a distinct theoretical approach (Hill and Wallace, 1996).

The aim of this book is to propose and demonstrate the utility of a multilevel theory of European foreign policy which goes beyond the analysis of the European Union. In particular, it seeks to illustrate that European foreign policy cannot be fully understood without consideration of Europe's relations with the United States and other international organizations engaged in Europe, such as the North Atlantic Treaty Organization (NATO), the Organization for Security and Cooperation in Europe (OSCE) or the United Nations (UN).

In exploring how such a multilevel theory of contemporary European foreign policy might be conceived, this book argues that this approach should meet several criteria. First, the theory should be able to analyse the increasing number and diversity of foreign policy actors engaged in European foreign policy. In order to do so, it should integrate the national, transnational and international levels of analysis. Second, it should recognise that key European foreign policy decisions are not only made by national governments, but increasingly also by international organizations. The latter might usually be preceded by national decisions, but the relationship between national and international policies is more complex with multilateral negotiations influencing national decisions and vice versa. Third, a theory of Europan foreign policy making should move beyond the description of European politics towards the explanation of processes and outcomes. Specifically, it should offer testable hypotheses concerning the behaviour of actors across levels of analysis and their impact on national and international policies.

In the following this chapter will suggest that the British policy network approach (Rhodes, 1986) in particular offers a suitable basis for the development of a multilevel theory of European foreign policy. Not only has this network approach emerged from the analysis of the relations between a variety of public and private actors, it has also been successfully applied to the study of decision-making at the national, transnational and international levels (Marsh, 1998; Benington and Harvey, 1998; Gummett and Reppy, 1990). However, this book proposes several modifications to the British policy network approach. In particular, it suggests a new way of combining the structural analysis of policy networks with rational choice theory.

The utility of the proposed multilevel network theory is subsequently examined in three case studies. The first case study analyses national and international foreign policy decision-making within the context of the European Union. Specifically, it studies the reduction of German controls for goods with civil and military applications (dual-use) during the European negotiations for common dual-use export regulations between 1992 and 1995. The second case study examines how European foreign policies are influenced by its relations within the transatlantic community, such as with the United States of America (US) and NATO. Specifically, it analyses the British decision to support the United States' drive for air strikes to protect UN safe havens in Bosnia in spring 1993. The final case study investigates to what degree the making of national defence policy is increasingly affected by the integrated multilevel European foreign policy network. In particular, this study examines the British abolition of its tactical air-to-surface missile (TASM) programme in 1993. In the conclusion, this book draws together the findings from the three case studies in order to assess the proposed multilevel network theory and its ability to provide new insights into the making of contemporary European foreign policy.

What is European Foreign Policy?

Before one can examine how foreign policies are made in contemporary Europe, it is necessary to define what is meant by European foreign policy. Typically European foreign policy has been understood as 'the sum of what the EU and its member states do in international relations' (Hill, 1998: 18). However, as the above has contended, European foreign policy cannot easily be reduced to the EU. Not only are the policies of EU member states and those of the European Union influenced by the United States and vice versa, but also there are key European foreign policy decisions taken and implemented by a broad range of national and multinational institutions, including the United Nations and NATO. This book, therefore, employs a different definition of European foreign policy which pertains to the decisions and actions of core European states and their multilateral organizations which are primarily concerned with the welfare of the region. The number of these core European states is constantly expanding. However, they are most easily defined by the overlapping membership of the European Union and NATO.

In addition, the question arises: what is meant by foreign policy? If contemporary foreign policy making is influenced by public and private actors at various levels of analysis, traditional notions of foreign policy as high politics, i.e. decisions involving heads of state, foreign secretaries and foreign office staff, are not sufficient to define the concept. Not only do a variety of actors participate in the decision-making process, but also the authority over the affairs among states has been increasingly transferred to organizations beyond national governments. It is therefore helpful to distinguish between foreign affairs and foreign policy decisions. While foreign affairs can be defined as the political deliberations and actions of public and private actors across national boundaries, the term foreign policy will be reserved here to denote authoritative political choices of action or legislative regulation at the national and international level which are directed to some actual or potential sphere outside the jurisdiction of the state polity (Kingdon, 1984).

The focus of this book on foreign policy decision-making derives from both empirical and theoretical concerns. Empirically, the emphasis on public decision-making builds on the observation that governments continue to hold a unique position with regard to the legitimate control over transnational and international affairs. In so far as international organizations have replaced them in determining authoritative political choices, they have done so on the basis of national policies - or the lack thereof (Mann, 1993). Indeed, most international organizations continue to subject themselves to the authority of national governments by providing member states with a veto - a feature which will be examined in more detail in the following chapters. Theoretically, the normative implications of the question to what degree and how European foreign policies are determined by national or international actors, places governments at the centre of this study. The aim of this book is not only to provide a better understanding of multilevel foreign policy processes, but also to provide an answer to the question of who controls European foreign policy.

Foreign Policy Decision-Making in the 1990s

The consensus that foreign policy decision-making cannot be adequately grasped by single-level analysis builds on a broad range of studies observing changes in the nature of the political process over the past decades. These studies widely agree that contemporary foreign policy decision-making processes in Europe and, more broadly, in the transatlantic community are characterized by three features: the increasing multiplicity, diversity and interdependence of foreign policy actors. It is difficult to assess the degree to which these three aspects have changed over the past decades. Interdependence among industrialized nations has been observed since the late 1960s (Cooper, 1972; Wallace, Wallace and Webb, 1977). At the time academics argued that the ability of governments to control their relations with other states was being curtailed by the economic integration associated with the emergence of multinational corporations and the European Economic Community. However, most scholars concluded that national governments maintained their decision-making power in the area of foreign and defence policy (Frankel, 1963; Keohane and Nye, 1989).

Since then much has changed. In particular the 1990s have seen a progressive transformation of European foreign policy decision-making. One feature has been the deepening and acceleration of the development towards greater multiplicity, diversity and interdependence of foreign policy actors in Europe and North America. In addition to economic developments, the end of the Cold War has led to greater integration in foreign and security policy. Expectations that foreign policies would be re-nationalized in the absence of the constraints of bipolarity (Mearsheimer, 1990; Waltz, 1993) have been disconfirmed. Contrary to the argument that European integration and the closeness of transatlantic relations relied on the specific conditions of the superpower competition, the EU and NATO not only continue to exist, but are in fact expanding their functional and geographical scope. The following examines each aspect in turn to illustrate that the trend towards greater multiplicity, diversity and interdependence among foreign policy actors at various levels appears has strengthened rather than reversed over the past decade.

Multiplicity

The notion of multiplicity commonly refers to the observation that the number of actors which are able to influence the foreign policy process and its outcomes has steadily grown over the past 50 years. Traditionally foreign and security policy appeared to be a distinct area of decision-making which predominantly involved heads of state, foreign and defence ministers and their respective ministries. Where the necessity arose to regulate transnational and international dealings, they were channelled through these two ministries. Today most governmental agencies within Europe conduct their daily foreign affairs directly with their respective counterparts in other countries. In the area of security policy, they are complemented by close formal and informal relations with the US through the UN, NATO and the OSCE as well as a large number of bilateral contacts.

However, the dispersion of influence in international relations has not been limited to governmental departments. Private actors directly participate in foreign policy decision-making because of transnational business interests or international causes, such as the protection of the environment and human rights. Transnational mergers have created an increasing number of multinational corporations which by means of their internal structure engage in international relations. Even in the armaments sector, national industries are increasingly the exception (Walker and Willett, 1993). Morever, non-governmental organizations have become regular actors in international relations. Valued as providers of information and services, as in the case of the International Red Cross, or feared as critics of governmental action, as in the case of Greenpeace or Amnesty International, non-governmental organizations have gained access to foreign policy making processes.

Furthermore, a range of international organizations has been created which function not only as fora for intergovernmental coordination but, due to their authority and staff, have often developed independent means and interests in international affairs. The density of these organizations in European foreign and security policy has increased steadily since the Second World War. It gained new impetus in the 1990s with the proliferation of international regimes and organizations in response to the perceived volatility generated by the end of the bipolar structure. Specifically, the North Atlantic Cooperation Council (NACC), its successor the Euro-Atlantic Partnership Council (EAPC) and the Partnership for Peace (PfP) were set up in order to establish security cooperation with Central and Eastern European states after the dissolution of the Warsaw Treaty Organization. Moreover, the functional and geographical scope of existing international organizations has been enlarged. NATO has been transformed from a collective defence organization into one of cooperative security. With or without the explicit mandate of the UN or the OSCE, the new NATO is able to conduct peacekeeping or peace-enforcing missions out-of-area, i.e. outside the territory of its member states. Moreover, at the 50th Anniversary of NATO on 16 March 1999, three former Warsaw Pact members, the Czech Republic, Hungary and Poland, joined NATO. Further applications for accession have been submitted by Albania, Bulgaria, Estonia, Latvia, Lithuania, Romania, Slovakia, Slovenia and the former Yugoslav Republic of Macedonia.

Similar developments have characterized the European Union and its former defence arm, the Western European Union (WEU), where many former Warsaw Pact members accepted associated partnerships. Furthermore, shortly after the NATO decision, the WEU too offered its resources for UN and OSCE out-of-area missions. In addition, Lithuania, Estonia and Latvia, Poland, the Czech Republic, Slovakia, Hungary, Romania, Bulgaria, Croatia and Slovenia as well as Cyprus, Malta and Turkey are seeking EU memberships. At the same time, the Conference for Security and Cooperation in Europe (CSCE) has developed from a forum for security negotiations into a regional organization under the UN charter, the OSCE. Its new tasks include amongst others the legitimization and monitoring of peace missions in the Euro-Atlantic area.

Diversity

The above enumeration indicates the second feature of European foreign policy: the actors involved are highly diverse. They not only cross the public-private divide, but also levels of analysis. The actors which participate in contemporary foreign policy decision processes are located at the national, transnational and international arenas. Although it can be argued that diverse actors have been engaged in foreign affairs at all times, the nature of their involvement has changed. As a consequence of functional differentiation within and across national borders, a broad range of actors have become affected by, and able to influence, authoritative decision-making with regard to foreign relations.

In particular, the taking on of governmental functions by private actors has increased their ability to influence foreign policies, not only in the area of trade, but also national and international security. Since the latter has been, until recently, a preserve of national governments, it shows specifically the new degree to which actors in foreign policy making have diversified. In the conflict in the former Yugoslavia, for instance, private actors participated as arms suppliers or mercenaries on the side of the warring factions and in the form of numerous charities which delivered humanitarian support while safeguarded by NATO troops on the side of the international peacekeeping mission. With their increased involvement in foreign relations, these actors also have growing influence over the European foreign policy decision-making process.

A similar transfer of governmental functions to the international level has increased diversity of foreign policy actors among international organizations. Thus, the end of the Cold War has seen a proliferation of new institutions which has enhanced the role of existing actors in foreign policy decision-making and introduced new ones. The transformation of the Conference of Security and Cooperation in Europe into an organization with a secretariat and permanent staff is one case; PfP and NACC are other examples. Moreover, after a period of perceived stagnation, the Maastricht and Amsterdam Treaties have significantly enlarged the authority of the EU not only in economic and monetary policy, but also in foreign relations with the establishment of the Common Foreign and Security Policy (CFSP) framework and, now, a European army.

The deepening of international institutions has been matched by a trend towards the widening of their memberships. Although the first candidates for EU accession have been Western European states, namely Austria, Finland and Sweden, many Central and Eastern European states have applied for accession. NATO has already accepted new members in Poland, the Czech Republic and Hungary and is set to invite more countries at its summit in Prague in 2002. Increasing diversity of foreign policy actors, therefore, can be noted within and without international organizations. Most notably within Europe, the widening towards Eastern Europe, has led to greater differences among the member states, and thus, regular actors in European foreign policy making, in economic, political, social and military terms. Externally, the institutionalization of international relations has created new organizational actors.

Interdependence

The result of the functional differentiation between governmental departments, public and private actors and international organizations described above has been increasing interdependence among a variety of foreign policy actors in the 1990s. Actors within and across national boundaries depend to a larger degree on each other's resources for the fulfilment of their needs and functions. Moreover, foreign policy decision-making and implementation has come to rely on contributions from a large number of actors.

In the private sector, increasing interdependence has been the result of specialization in production on one hand and global marketing on the other. In the public sector, governments have increasingly been willing to accept the interdependence that comes with multinational economic and political collaboration as well as public-private partnerships. A particular example in the area of foreign policy making has been the growth of cooperation in armaments research and development, which has previously been identified with national sovereignty. Not only have national armaments industries lost their military rationale if national defence and international interventions depend on the cooperation of allies, but it has also become more difficult to defend high military spending politically and economically if it is less costly to buy weapons from allied countries or to collaborate in arms production. However, as governments sell national armaments industries to private actors, accept transnational mergers of procurement companies and favour international cooperation in the development and production of weapons, national defence policy becomes more vulnerable to transnational and domestic influence (Guay, 1998; Wulf, 1993).

In addition, transnational and international interdependence has increased as a result of the functional and regional enlargement of international organizations. In particular in the area of security policy, interdependence has reached new levels in recent years. One reason for this has been the progressive decline in national defence budgets. Since the end of the Cold War, popular demands for a peace dividend have further reduced national defence capabilities to the degree that large scale interventions and national defence rely on multilateral cooperation (Smith, 1993). International peacekeeping missions, such as in the former Yugoslavia or in Afghanistan, thus typically involve troops from more than ten countries.

Multilevel Approaches in Foreign Policy Analysis

As a consequence of the transformation of the foreign policy decision-making in Europe and North America, a number of theories have been developed which seek to analyse the growing interconnectedness among foreign policy actors at different levels. In particular, three approaches should be noted: transnationalism, the two-level game and network models. Each approach has offered new insights into the multilevel nature of foreign policy making. However, none of the models integrates all three levels of analysis. This section discusses the advantages and disadvantages of each model in turn before suggesting that the British policy network approach might be particularly suited for the development of a multilevel theory of European foreign policy making.

Transnationalism

Transnationalism is perhaps the oldest multilevel approach to international relations and foreign policy analysis. It originated in the 1970s from the theoretical work of, among others, James N. Rosenau (1969), Robert O. Keohane and Joseph S. Nye (1989). In 1995, Thomas Risse-Kappen sought to revive the interest in transnationalism with the volume '*Bringing Transnationalism Back In*'.

Transnationalism investigates how cross-border relations help public or private actors to influence foreign policies abroad. Analytically, two forms of transnational interaction can be distinguished. Trans*governmental* relations refer to the actions of state agents, such as politicians and civil servants, which are in regular contact with actors in other nations. Trans*societal* relations denote the growing networks among private actors from different countries. A large number of structures and processes have been identified which shape these relations. In one of the earliest frameworks for the analysis of national-international linkages James N. Rosenau (1969: 52) suggests no less than 24 dimensions for the study of transnational relations, including the type of actors involved, their attitudes, the institutions which define their political system and the processes of their interaction. In a more recent study, Thomas Risse-Kappen (1995) argues that transnational relations are defined by the political structure of the state, i.e. its political institutions, the cohesiveness of its society and the network of social relations, as well as international structures, such as regimes.

However, in spite of the broad variety of factors analysed by transnationalist approaches, they provide only a selective understanding of contemporary foreign policy making. Most crucially, transnationalism remains essentially a single-level theory. As Risse-Kappen points out, transnationalism focusses exclusively on transnational relations and therefore needs to be combined with theories of national and international foreign policy in order to provide a comprehensive analysis of contemporary foreign policy making. Moreover, transnationalism rarely explains particular foreign policy outcomes. Although a number of hypotheses have been proposed by different transnationalist approaches (Risse-Kappen, 1995; Evangelista, 1995), these hypotheses do typically not go beyond the explanation of access to foreign policy decision-makers or foreign policy processes. How this access enables transnational actors to influence national or international foreign policies can only be explained with additional theoretical models.

Two-Level Games

A second multilevel approach can be found in the two-level game by Robert Putnam (1988). Next to transnationalism Robert Putnam's two-level game has perhaps gained the most widespread recognition in the mainstream multilevel analysis of international relations. Since its publication, the approach has been employed in a wide range of studies, including an edition published by Peter Evans, Harold Jacobson and Robert Putnam (1993) which has sought to further refine the model.

The two-level game explains intergovernmental negotiations in terms of a two-stage game in which diplomats simultaneously seek to accommodate domestic and international demands. These demands are analysed on one hand in the form of different domestic coalitions, termed win-sets, which support a certain political outcome, and on the other by the preferences of the second party in the international negotiations. The two-level game explains negotiation outcomes as the result of the strategic interactions between the two negotiators. Specifically, the negotiators, who are believed to hold a gatekeeper position between the national and the international arena, can coax their domestic audiences into accepting an agreement by pretending that the constraints of the opposing side will not allow for any compromises. The negotiator can also extract higher international concessions by alleging strong domestic pressures.

In spite of the interesting insights offered by the two-level game into the strategic nature of multilevel decisions, several criticisms can be made with regard to the synthesis of the different levels and its analysis of the decision-making process. Notably, the original model, specifically its notion of governmental negotiators as gatekeepers, excludes the transnational level of analysis. Jeffrey Knopf (1993) has sought to addressed this shortcoming with his development of a three-level game approach. However, Knopf's integration of transnational actors continues to omit international organizations, such as the EU or the OSCE, who frequently mediate in intergovernmental negotiations, nor does it analyse how third states can directly or indirectly influence bilateral foreign policies. Furthermore, due to its primary concern with the level of the international negotiations, the two-level game does not analyse whether or how domestic actors are engaged in the foreign policy decision-making process, be it by actively seeking to influence their or the other negotiator or by attempting to enlarge the support for their policy preferences among other national actors.

Policy Network Analysis

The perhaps best basis for an integration of the national, transnational and international level in international relations is provided by the British policy network approach. Although the British policy network approach was originally developed to examine the relations between public and private actors in domestic policy making (Rhodes, 1986), it has increasingly been applied to the analysis of transnational and international relations in Europe (Peterson, 1992; Benington and Harvey, 1998). American and German network analysis has conversely remained very much wedded to the domestic domain (Scharpf, 1988; Knoke, 1990).

Crucially, since the British approach defines networks as all actors who share an interest in a specific issue area and who are linked to each other through stable formal or informal relationships (Atkinson and Coleman, 1992), they can include domestic, transnational and international linkages. Moreover, policy network approaches allow for a broad variety of relations, ranging from hierarchical to pluralist or horizontal arrangements (Kenis and Schneider, 1991: 42), which make the network concept especially suited for theorizing about decision-making structures at different levels of analysis. The key hypothesis of the policy network approach is that the distribution of contacts among public and private actors and the nature of their relations, i.e. the structure of a network, determines the ability of its members to influence decision-making processes (Waarden, 1992). Thus, various network models suggest different types of networks in which certain actors or coalitions of actors typically determine the outcome of the decision-making processes.

Nevertheless, for the purposes of analysing foreign policy decision-making in contemporary Europe and North America, the British network approach is challenged in two ways. First, although policy network models have been employed in empirical studies of decision-making at different levels, little attempt has been made to theorize about the synthesis of multiple levels within the network approach. Second, few British network models help to explain the decision-making process in terms of the formation or changes in the coalitions which define policy outcomes. In fact, due to the predominant concern of the British network approach with the link between network structures and policy outcomes, most models neglect the role of decision-making processes as an intermediate variable. Typically, network models offer typologies which assume the form of quasi-causal propositions. As Maurice Wright (1988) observes, policies are more often read off a type of network than explained.

However, several suggestions have been made to improve the British network approach. In particular, the limitations of network typologies for the explanation of the decision-making process have been widely recognized. In response, networks are increasingly treated as unique, and analysis has focussed on the mapping of individual networks rather than the categorization into ideal types. In addition, it has been suggested that network analysis can be fruitfully combined with rational choice assumptions in order to hypothesize about changing policy preferences and coalitions. The following examines in detail how these suggestions can be used to transform the British policy network approach into a multilevel network theory of contemporary foreign policy making.

Towards a Multilevel Network Theory

Key for modifying the British network approach into a model consistent with multilevel analysis lies in the development of a theoretical definition of networks which explicitly includes national, transnational and international actors. Furthermore, the basic concepts and dimensions of the policy network approach have to be reexamined with regard to the question whether and how they can be consistently applied to multiple levels. In doing so multilevel network theory builds upon the contention advanced in this chapter that the distinction between domestic, transnational and international policy-making, which has so far shaped network analysis, has been superceded by changes in the decision-making process in Europe and the transatlantic community. Not only are actors linked across levels of analysis, the structure of their relations in the domestic and international arena has also become increasingly similar. The traditional ideal-typical distinction between hierarchical, institutionalized relations in the national system, and anarchic, informal linkages in the international system which underpinned the division between theories of international relations and foreign policy analysis meets empirical observations less and less. Both in the domestic and the international arena, we find today a mixture of formal and informal relations between the public and private actors engaged in foreign policy decision-making. The following chapter therefore suggests that these formal and informal relations can be defined by a single concept of power based on a combination of resource-dependencies and institutional authority.

In addition, the subsequent chapter examines how the British policy network approach can be combined with rational choice theory in order to hypothesize about the decision-making process as an intermediate variable between structures and outcomes (Dowding, 1994: 60; Daugbjerg and Marsh, 1998: 67). However, it uses a different concept of rational choice than German or American network models which have attempted the synthesis of network analysis with game theory (Scharpf, 1988). The following suggests specifically that the utility of rational choice theory for network analysis derives from its ability to explain how rational actors may use their relations in a network to exert pressure on each other and thus understand the interactions among the members of a network. In particular, this author argues that the concept of *bounded* rationality conforms with the British network approach in that the members of a network by definition interact regularly with each other. It can, therefore, be presumed that actors have clear expectations regarding the cost and utility of exerting pressure on other actors in order to change their policy preferences within their network. In short, the structure of a network sets the boundaries in which actors rationally seek to influence the decision-making process.

Moreover, the rational use of network relations, if defined by resource-dependence and institutional authority, has been indirectly pointed out by studies which have investigated the compatibility of rational choice theory with new institutionalism or behaviouralist analyses (Dowding, 1994; Nørgaard, 1996; Searing, 1991). The advantage of rational choice assumptions for the study of political decision-making processes is that they provide general hypotheses about the ways in which actors use their relations to influence each other and the decision-making process. By doing so, rational choice assumptions cannot only be utilized to illustrate some common features in the actions of a variety of actors, but also to generalize interactions across cases and issues. While the focus on the commonalities of the decision-making process regarding different issues necessarily limits the understanding of a particular historic foreign policy decision, it permits comparisons and the development of general propositions which might help to explain other cases.

Finally, the following chapter proposes a quantitative measurement for the ability of actors to influence each other in terms of the number of actors in the network who exert pressure on a single actor at any moment of the decision-making process. Multilevel network theory, thus, addresses another criticism of British policy network models, namely that they are able to observe and describe actors' influence over policy outcomes, but do not offer testable hypotheses which help to explain the process and dynamics of influencing policy outcomes (Dowding, 1998; Peters, 1998:24). The proposed quantitative measurement reflects the notion advanced by, amongst others, pluralist and corporatist approaches, that political actors, such as ministers or even civil servants, are responsive to the policy preferences of their constituencies, understood here in the wider sense as the actors on whom they depend and with whom they regularly interact. Moreover, analysing the number of actors who support a policy is implicitly linked to the ideal notion of democratic decision-making in that both elections and parliamentary decisions are commonly based on the preferences of a majority.

Theory Testing and Case Selection

The subsequent testing of multilevel network theory in three case studies concerning European foreign policy making within different national and international contexts raises three questions: How many case studies have to be conducted for a valid assessment of the theory? What defines suitable case studies? When are the theory's hypothesis corroborated?

The first question is associated with the problem of 'many variables, small N' which means that the number of explanatory variables exceeds the number of cases (N). In these circumstances, the test of a theory will always be inconclusive (Lijphart, 1971). Fortunately, within the context of this book the problem could be avoided since the hypotheses which will be proposed in the following chapter concern the preferences of each actor within the network and not merely the outcome of the decision-making process. In fact, each of the three cases presented in this book has several outcomes, depending upon whether the reader is primarily concerned with the decisions of the EU or Germany in the first case study or the United Kingdom, the United Nations or NATO in the second, for instance. Moreover, each of these decisions was preceded by a series of preference changes among other actors within the multilevel European foreign policy network which contributed to the formation of a winning coalition in favour of a particular foreign policy. Thus, instead of the number of case studies, the number of preference changes were crucial for the test of multilevel network theory. Each instance in which a member of the network changed or maintained his or her policy preference after other network actors had modified their views concerning a particular foreign policy represented a test case for the hypotheses.

The second question concerns the difference between independent and dependent variables and the scope of the proposed theory. Both necessitate different conditions for the selection of appropriate case studies. In order to assess the ability of the independent variables to explain changes in the dependent variables, possible alternative explanatory factors should be held constant. To demonstrate the general applicability and transferability of the proposed hypotheses, the theory should be applied to cases which are as diverse as possible with regard to the independent variables (Mackie and Marsh, 1995:176). As a result, the cases presented in this book were on the one hand selected with the aim to reduce variation in the multilevel European foreign policy network as far as possible. On the other hand, this book sought to maximise the variation between the cases in order to assess the utility of multilevel network theory for the explanation of the European foreign policy making in different institutional and situational settings.

To meet the first criteria, the cases were drawn from a comparatively narrow period of five years between 1990 and 1995 which followed the main transformations associated with the end of the Cold War and preceded the enlargement of NATO. Certain changes in the relations among political actors in the multilevel European foreign policy network, however, could not be avoided because historical case studies prevent the full control of exogenous factors or because these changes occur routinely as part of the political process. Specifically, elections have the potential to relocate political actors in different institutional positions within the network, had to be reflected by changes in the map of the multilevel European foreign policy network presented in the appendix.

To meet the second criteria, this book considered three settings: the European Union, the transatlantic community and national foreign policy making. In addition, it selected cases which covered a diversity of issues ranging from economic relations to international security and nuclear defence. Finally, this book attempted to apply multilevel network theory to the analysis of European foreign policy decisions at different levels. Thus, each case study examined the decision-making processes which defined the foreign policies of one European country, the main international organizations involved and the influence of the United States on European foreign policy and vice versa. The focus on only one European state in each case study appeared necessary to reduce the scope of the analysis, both with regard to the detailed mapping of the multilevel European foreign policy network and the examination of the decision-making process which typically stretched over one to two years. However, incidentally Germany and Britain appeared to be particularly interesting examples in the three selected cases because of unexpected changes in their foreign policies. Moreover, both countries play central roles in European foreign policy both through their individual capabilities and contribution to international organizations.

The question when the proposed hypothesis might be regarded as corroborated is the most difficult to answer, since the hypotheses proposed by multilevel network theory are necessarily probabilistic. As such, they prevent their assessment through simple falsification (Gillies, 1971; Lakatos, 1970). Deterministic causal relations which are the basis for falsifiable hypotheses, however, can only be rejected in social science. Not only is it impossible to control for the full range of environmental factors which might affect a single case study, the nature of social interaction itself prohibits a deterministic conception of causality since the social world is by definition inter-subjectively constructed.

The empiricist study of international relations, nevertheless, asserts that theoretical explanations of social phenomena can be assessed. Several criteria have been agreed upon as evaluative standards for theoretical models within this methodological framework, such as explanatory power, progressive research programme, consistency, parsimony and their correspondence with empirical findings (Young, 1972; Vasquez, 1995; Most and Starr, 1980). These criteria imply that theories and alternative hypotheses have to be evaluated in comparison to each other. The following empirical studies seek to enable such a comparison by measuring the number of instances in which the hypotheses were confirmed. Finally, the conclusion of this book discusses the new insights offered by multilevel network theory into the nature of contemporary European foreign policy in comparison with the three multilevel theories outlined in this chapter. Whether the degree to which the hypotheses meet empirical observation is acceptable and the contribution of multilevel network theory to our understanding of foreign policy decision-making is unique will have to be decided in the academic debate.

Chapter Outline

The findings of this study are presented in five chapters. The first chapter develops a multilevel network theory based on the suggestions made above. This chapter examines how the key concepts of the network approach might be redefined for multilevel analysis and how the approach can be combined with rational choice theory in order to explain decision-making processes. The subsequent chapters proceed to test the utility of multilevel network theory for the explanation of national and international foreign policies within the context of the EU, the transatlantic community and national security. Specifically, the analysis of foreign policy decision-making within the EU examines the European agreement on common export controls for technology with civil and military applications between 1992 and 1995 and the resistance of the German government to a reduction of its controls in its wake. The chapter on foreign policy making in the transatlantic community analyses the decision of the UN Security Council and NATO in favour of air strikes in Bosnia in 1993 and the British foreign policy u-turn which paved the way for the endorsement of air strikes. And the chapter on national foreign policy making studies the decision of the British government to abandon its tactical air-to-surface missile programme. The conclusion of this book draws together the findings from all chapters for a general evaluation of multilevel network theory and to summarize its insights into contemporary European foreign policy.

2   Multilevel Network Theory

Introduction

This chapter modifies the British policy network approach (Rhodes, 1986; Rhodes and Marsh, 1992) in order to enable it to model multilevel foreign policy decision-making processes in Europe and North America and to enhance the explanatory capabilities of this approach. In particular, the following sections address four questions. First, how should networks be defined in the context of the increasing multiplicity of actors engaged in foreign policy making. Second, what are the consequences of the growing diversity and interdependence of network actors for the analysis of power relations in networks. Third, what does the increasing variety of foreign policy actors mean for the conceptualization of political agents in multilevel networks. And fourth, how can rational choice theory help to explain changes in the foreign policy preferences of network actors and the formation of a winning coalition during the foreign policy decision-making process. In conclusion, this chapter outlines the operationalization of the model in the subsequent case studies.

Concepts and Definitions

It has been argued in the previous chapter that the concept of policy networks is especially suited for the analysis of contemporary multilevel decision-making because it is able to model a variety of actors engaged in the modern policy process as well as the flexible and multifaceted relations among them (Bressers et al., 1994: 5; Kenis and Schneider, 1991: 42). This ability is based on a definition of policy networks as a set of public and private actors who share an interest in a particular issue area, who routinely interact with each other and who are connected to each other through stable formal and informal relations (Atkinson and Coleman, 1992). The problem which arises from this definition, however, is that it provides no clear criteria for delineating the boundaries of distinct networks. The attempt to define the boundaries of networks has therefore led to various approaches in network analysis. Most network theorists have attempted to distinguish separate networks by their internal features. Thus, according to R.A.W. Rhodes and David Marsh (1992), different networks are determined by their membership, degree of integration, distribution of resource dependence and distribution of power. In a review of different network models, Frans van Waarden (1992) observes that networks are commonly distinguished along seven dimensions. They include the number and type of actors, function and structure of the networks, the degree of institutionalization, rules of conduct, power relations and actors' strategies.

A close scrutiny of these dimensions shows that distinct networks are essentially defined by two aspects: their agents and their structure. In addition to the above mentioned disagreements over which specific dimensions should be taken into account in delineating different networks, this poses a particular problem. According to the definition of networks, the characteristic feature of networks is that they include a diversity of agents and different types of relations. Moreover, each network combines a different mixture of them. It follows that attempts to delineate the boundaries of networks on the basis of a distinction, for instance, between public and private actors or hierarchical and horizontal structures are inherently inconsistent with the notion of networks. Returning to the definition of networks, it emerges that the only characteristic which distinguishes one network from another is the stability of the relations among the actors and the regularity with which they interact. It follows that the boundaries of networks can only be identified as disconnections among sets of political actors. J.K. Benson has pointed this out in an early sociological definition of a network as a 'complex of organizations connected to each other by resource dependencies and distinguished from other ... complexes by breaks in the structure of resource dependencies' (cited in Rhodes, 1988: 77).

Network Boundaries

Two lines of reasoning support the proposition that these breaks generally comply with policy sectors or issue areas, such as education, health, agriculture, transportation, monetary policy, energy or labour (Wright, 1988: 596; Atkinson and Coleman, 1992: 157; Heinz et al., 1990). The first argument contends that networks conform with the sectoral division of policy sectors because of functional differentiation among political actors (Marin and Mayntz, 1991: 17). It proceeds from the observation that political decision-making in European democracies is structured along the divisions of labour between sectoral ministries. Separate departments deal with policy making and implementation in agriculture, health or defence, for instance. Moreover, the sectoral division of public institutions shapes the relations through which private actors can seek to influence the political decision-making process. Thus, large armaments companies will usually have strong and stable ties with ministries of defence, while farming associations are typically linked to ministries for agriculture.

The second argument in favour of the sectoral boundaries of networks is that stable relationships evolve among actors who depend on each other for the exchange of material or ideational resources, such as money or expertise. Since political influence and information are ideational resources, the second argument supports the first. In addition, it points out the role of resource-dependencies in defining relations which are not institutionalized, but informal and flexible. Expressions of such relations in the private sphere include the subcontracting of production as well as collaboration in research and development among companies in the same sector. In the multilevel European foreign policy network, such informal relations can be found in particular in Britain where industry relations were deregulated in the 1980s. In Germany, conversely, existing resource-dependencies and commonalties of interest have produced strong industry associations which institutionalize the relations within sectors and represent their members vis-à-vis governmental actors.

Foreign Policy in Multilevel Networks

Foreign policy analysis has traditionally posed a problem for the delineation of network boundaries because it did not conform with the sectoral divisions of domestic policy processes. On the one hand foreign policy making transgressed national boundaries because it routinely involved transnational and international actors, on the other hand it crossed sectoral lines because each ministry conducted its foreign relations through the foreign affairs department. The transformations of the European foreign policy decision-making process described in this book have changed both conditions (Marsh and Rhodes, 1992: 258; Peterson, 1992; Gummett and Reppy, 1990; Atkinson and Coleman, 1992: 163). Transgovernmental relations today conform with issue areas since departmental ministries increasingly cooperate directly with their counterparts in other European countries and across the Atlantic. In fact, sectoral departments such as the economic ministries now often take the prime responsibility for leading EU negotiations in their issue area. Functional divisions also dominate among international organizations which channel transnational and international cooperation between states. Moreover, where international organizations have attained some authority over foreign relations, these have been structured along functional lines. Indeed, the European Union itself is structured according to issue areas which are represented by the divisions within the Commission, sectoral councils and their hierarchies of committees. It follows that rather than adapting the network concept for the analysis of multilevel foreign policy decision-making, recent changes of the European foreign policy process have increased its similarity with domestic decision-making and thus made it more susceptible to network analysis.

The consequence of the expansion of sectoral decision-making structures in virtually all issue areas across national boundaries, however, raises the question whether a distinct foreign policy network exists. If most sectoral policy networks in contemporary Europe cross levels of analysis, they all are potentially foreign policy networks in that some of the decisions made within them would be directed to some actual or potential sphere outside the jurisdiction of the state polity (Kingdon, 1984). Nevertheless, it can be argued that a foreign policy network can be found alongside increasingly transnational sectoral networks such as agriculture, industry or telecommunications. Although the latter are characterized by transnational and transgovernmental interaction in political decision-making, the decisions taken within these networks typically apply to and are implemented at the domestic level.

As such, these decisions are not foreign policies as in the introduction of this book. Furthermore, based on the definition of networks as sets of actors who share a specific interest in a particular issue area and who are linked to each other through stable relations, a range of actors networks can be identified which are predominantly concerned with policies directed to and implemented at the international arena. They subsume all public and private actors whose primary interests lie beyond the national boundaries of their countries. Actors concerned with national security and defence policy certainly fall into this category as their only interests are matters beyond their national boundaries. However, foreign policy networks also include export industries because they are mainly affected by the political regulation of transnational relations.

Finally, the question emerges how networks in general, and foreign policy networks in particular, change over time and how this affects their delineation. The previous chapter has suggested that the European foreign policy decision-making process has been considerably transformed over the past 40 years. This transformation has not only included the emergence of new actors, but also the increasing interdependence between public and private actors at the national and international level. Analytically two causes can be distinguished which may lead to changes in networks: external and internal causes. External causes might involve the advent of new actors who share an interest in an issue area and seek to gain access to the decision-making process. Internal causes may be self-induced changes, e.g. political decisions which redefine the formal relations among sets of actors in a sector, material changes in the resources of actors, or routine and institutionalized changes, such as elections which bring new parties into government. All impact directly on the delineation of networks and, therefore, any map of a network can only provide a temporary snapshot of the actors and relations involved. However, since a network by definition consists of a stable set of agents who regularly interact, only long-term changes which apply not only to a single case, but indicate a more permanent transformation of the network across a range of issues, are considered relevant. Within the five-year range of the cases examined in this book, the institutional changes in a number of international organizations, such as NATO and the OSCE, certainly belonged to this category. But elections, such as the coming to power of Bill Clinton in the US, also changed some relations within the European foreign policy network. The following section examines how these relations are defined.

Power in Multilevel Networks

While the actors and their relations cannot distinguish between networks, the two are essential for the analysis of network structures and the ways in which they affect the decision-making process. Since multilevel network theory seeks to explain how political actors influence each other's policy preferences, this structure has to be analysed in terms of how it affects the ability of actors to exert influence. The ability to influence is grasped in the concept of power. However, the notion of power has been regarded as an essentially contested concept (Connolly, 1983). In particular, different conceptions and measures of power have been employed in the analysis of the domestic on one hand and the international policy making on the other. This section develops a definition of power which can be applied to actors across multiple levels of analysis.

Power as Causal Hypothesis

A theory which seeks to analyse the ability of political actors to influence each other is *per definitionem* based on a concept of power. Power can be defined as the potential or actual ability of an actor A to deliberately change the preferences or the behaviour of another actor B with respect to an issue X (Dowding, 1991: 68; Connolly, 1983; Baldwin, 1989). It can be differentiated from the concept of influence in that power can be both the potential and actual capacity to modify another actor's beliefs or actions, while influence only applies to actual, observable changes. Two key features characterize this definition of power. First, it describes a dyadic relationship, namely the relation between two actors A and B (Emerson, 1962; Baldwin, 1989). It follows that the analysis of networks which utilizes the concept of power should describe a network as a two-dimensional matrix of dyadic relationships as in Table 2.1.

Table 2.1   Matrix of dyadic power relations

	Actor A	**Actor B**	**Actor C**	**Actor D**
Actor A	-	relation A to B	relation A to C	relation A to D
Actor B	relation B to A	-	relation B to C	relation B to D
Actor C	relation C to A	relation C to B	-	relation C to D
Actor D	relation D to A	relation D to B	relation D to C	-

The second feature of this definition of power is that, unlike influence, it cannot be directly observed or measured because it also denotes the potential to influence. Power has to be analysed either deductively or inductively, i.e. it can be inferred from A's capabilities in advance of its exertion or it can be measured *a posteriori* by changes in B's behaviour or preferences (Connolly, 1983). While the former relies on the study of material and ideational properties, the latter investigates behaviour. Both methods have been used in the study of domestic policy networks. However, only the deductive approach suggests an explanation as to why the actor A can influence actor B. Conversely, the inductive approach concludes that power is the result of influence. Where an actor A has the observed ability to influence B, it is presumed that he or she has also the capability to do so in future. The question what enables A to modify B's behaviour is not addressed.

The difference between the two approaches is based on their understanding of the nature of power. The inductive approach treats power as a type of relationship, while the deductive approach conceives of power as a causal hypothesis (Nagel, 1975). Specifically, the deductive approach suggests that a causal relationship exists between the preferences and actions of actor A and the preferences and actions of actor B. However, in order to denote a causal relation and not merely a correlation, the deductive approach has to distinguish between cause and effect. Stating that A's and B's preferences and actions correlate would not describe a genuine causal relationship between both actors. A causal relationship has to identify the direction of the causality, i.e. it has to differentiate whether A influences B or vice versa. Analytically, four types of power relations can be distinguished with regard to causation, i.e. the direction in which power can be exerted: (1) A has power over B [A>B], (2) B has power over A [B>A], (3) A and B have power over each other or, more succinctly, A and B are interdependent [A<>B] and, (4) neither A nor B has power over the other or A and B are autonomous [A|B]. The structure of a network can then be described as the distribution of these four types of relations among all members of the network.

The Bases of Power

In addition to distinguishing between the direction of a causal power relationship, the distinction between cause and effect has to be justified in order to meet the definition of causality. Such justification is provided by an explanation as to *why* A has the actual or potential ability to influence the behaviour and preferences of B. In political theory such an explanation is referred to as the bases of a power relation (Connolly, 1983). Two alternative explanations have dominated the analysis of domestic policy analysis on one hand and international relations on the other until the 1980s. In international relations, power was traditionally associated with material resources, while in the domestic system power was predominantly explained by institutional structures (Rhodes, 1986: 17). The different explanations were a result of the perceived structural differences between the national and the international system. Thus, in the domestic arena, institutional analysis traditionally featured strongly because of the perceived dominance of formal, institutionalized decision-making structures. The international arena, which was viewed by neo-realists as the realm of anarchy characterized by the absence or limited influence of formal institutions, material resources were favoured as indicators of power (Waltz, 1979; Keohane, 1984).

　　Although the distinction between institutionalized domestic structures and international anarchy has never been clear-cut, the transformation of decision-making processes in the transatlantic community has led to an increasing convergence of domestic and international structures. While domestic public-private relations have been characterized by deregulation since the 1980s and privatization continues in sectors such as telecommunication, transport and health, international institutions have proliferated due to the expansion of the functional scope of the EU, NATO or the OSCE for instance. As a consequence of these changes and due to the extension of decision-making networks across systemic boundaries, it has become necessary to integrate the analysis of resources and institutions for the study of power in multilevel networks. The following investigates how the analysis of institutions and resources can be combined across levels of analysis in order to deduce power relations in multilevel networks.

Institutions

Since this book defines power as a causal relation, its analysis requires relational indicators. This does not pose a problem with regard to institutions. Institutions denote social relations in their definition as 'persistent and connected sets of rules (informal and formal) that prescribe roles, constrain activities and shape expectations' (Keohane, 1989: 3). They attribute competence and legitimate authority to certain political actors within the decision-making process by describing their relations with other actors and by prescribing legitimate modes of action between actors. In Europe, most formal institutional relations are codified in national constitutions, laws and regulations. In addition, informal institutional relations have emerged through convention and can be observed in the regular interactions between public and private actors. Britain holds a special position in this respect because it does not have a written constitution, but relies primarily on conventions for the definition of its institutional relations in the political realm. However, as has been argued above, formal institutional relations are not confined to the domestic level. In the international arena, formal institutional relations have been set up by treaties, regimes and documents of the main international organizations. They not only define legitimate relations and modes of interaction among state governments, but also between private actors, such as firms, interest groups or even individuals.

Resource-Dependence

Contrary to institutions, the distribution of resources among actors at different levels of analysis may be coined in relative terms, but it is not a relational concept. An actor's possession of specific resources does not *per se* reveal anything about his or her power relations with other actors. Although an actor might attempt to use his or her resources in bargaining situations or to force other actors to modify their behaviour, these efforts are likely to be unsuccessful if the targeted actors have control over similar resources. In order to provide A with power over B, A's resources have to meet the needs and/or lack of resources by B (Emerson, 1962; Keohane, 1989). Specifically, resources can be the basis of two forms of power relations. First, resources can be used in exchanges as described by the relational concept of resource-dependence among actors. Robert O. Keohane and Joseph Nye define such (inter)dependence as situations 'where there are reciprocal (although not necessarily symmetrical) costly effects of transactions' (Keohane and Nye, 1989: 9). Second, resources can be employed to force or threaten to force actors to change their behaviour if they lack matching resources.

Since multilevel networks are based on a stable set of actors who regularly interact with each other, it can be argued that physical force or the threat of force plays a negligible role in the analysis of multilevel networks. As will be argued in more detail below, all network actors potentially depend on each other for the formation of coalitions in the political decision-making process. The use or threat of force would endanger future cooperation in such coalitions and, therefore, is believed to entail greater long-term risks than short-term benefits. As a consequence, resource-dependence has been regarded as the primary basis of material relations within networks. They require the examination of two variables: the distribution of resources and the respective needs of each actor. Given the variety of actors in multilevel networks, multilevel network theory practically rules out the possibility of arriving at a conclusive list of power resources. Conversely, multilevel network theory requires a flexible approach to power resources which essentially includes all tangible and intangible assets which may determine resource-dependence relations between any two particular actors. Due to the infinite range of resources which may be the basis of resource-dependence relations, the analysis of the exchange relations among network actors best begins with an examination of the specific needs of each actor and by whom these needs can be met.

In multilevel network analysis needs can best be defined as the 'objective welfare demands' of actors because they allow needs to be deduced from the basic functions of the actors (Dowding, 1991: 35). According to this definition, needs stem from basic physical requirements necessary to ensure survival and prosperity. Commonly, physical needs of individuals are listed as food, shelter, safety and employment (Oppenheim, 1981: 128, 141). They also include all resources which are required for the fulfilment of the specific functions of public and private actors, such as the development and implementation of policies by ministerial departments and the provision of goods and services by private companies. Most of the functions of public actors are laid down in the documents which define their institutional relations with other actors within the network, while the functions of private actors arise from the objective demands of organizational welfare, i.e. the survival of a firm in a competitive market, and the need for resources for production and service. Once an actor's functional need for a specific set of resources has been established, it can be analysed which actors in the network are able to meet these needs and stand in a resource-exchange relation with the actor.

Combined Power Structure

While the above asserts that institutional and resource-dependence relations are defined by different variables, both simultaneously define the type of power relation between any two actors in the network. However, the type and direction of power as determined by the two dimensions can differ. Thus, the resource-dependence relationship between actor A and actor B may give B power over A, while the institutional relationship between both might give A power over B. An example would be the relationship between a minister and his or her civil servants. Although the minister depends on information and expertise from the civil servants, he or she has institutional authority over their actions. In order to understand the power relations between any two actors in a multilevel network, therefore, the combined effect of both dimensions on the ability of actors to influence another has to be analysed.

The cumulative influence of institutions and resource-dependencies on power relations can be perceived as a two-dimensional space. In each dimension, the power relation can take one of the four types of causal direction identified above, namely A has power over B [A>B], B has power over A [B>A], A and B are interdependent [A<>B] and A and B are autonomous from each other [A|B]. However, in so far as resources and institutions determine the relationship between two specific actors in a single network, they generate one power relation between them which combines both.

The cumulative definition of the power relation between any two actors by the two dimensions can be understood by three logical axioms. First, if one dimension is characterized by interdependence, the combined power relation is also interdependent. This proposition can be justified because the mutual dependence of the actors on each other cannot be terminated by any other type of relationship in the other dimension. To illustrate: If two ministries depend on each other for the exchange of information and expertise, a higher institutional authority of one ministry over the other in some issues does not change the fact that they are mutually dependent. From this axiom follows, second, relations where either A or B has power over the other only exist where the institutional and resource-dependence dimensions are characterized by either power relations with the same direction or if one relation is marked by autonomy. Thus to stay with the above example, the power relation between two ministries would be described as A has power over B [A>B] if ministry B was dependent upon resources from ministry A [A>B] and institutionally subordinate [A>B] or autonomous from it [A½B]. Third, if one dimension is defined as A has power over B [A>B] and the other as B has power over A [B>A] , the resulting power relation can be defined as interdependent [A<>B]. For instance, it can be argued that if actor A depends upon the resources of B, but can influence B because of his or her institutional authority, both actors will have the capability to exert power over another. To take another example from ministerial relations, such a combination is represented by the relationship between ministers and their civil servants. Typically, ministers would have the superior institutional control over the bureaucratic apparatus, but they require expertise and information which are provided by the civil servants. As a consequence both can exert some power over each other.

Direction versus Degrees of Power

The additive axioms presented above neglect that power can be symmetrical as well as asymmetrical (Oppenheim, 1981: 34). The analogy of resource-dependence with supply and demand relations points to the fact that the power relation between any two actors A and B cannot be accurately understood outside the context of their respective relations with other network actors. A's power over B might be diminished, if B is able to satisfy his or her needs from alternative sources, such as actors C or D. Viewed from this perspective, power relations are not absolute as implied by the four types of power relations identified above. Depending on the availability of resources, power should rather be conceptualized as a continuum which allows for different degrees. The same argument can be made with regard to institutions. Some actors have higher institutional authority over a particular actor than others. For instance, although both a parliamentary political party and a minister have the ability to influence the prime minister, the institutional influence of the minister will commonly be regarded as stronger than that of the parliamentary party.

In spite of the apparent reductionism of distinguishing merely four types of power relations in terms of their causal direction, several arguments support the usage of this approach for multilevel network theory. They show that a directional typology of power relations is not only empirically more rigorous than degrees of power, but also more conducive to the network approach. The main problem of degrees of power lie in their theoretical conceptualization and empirical measurement if power is defined in relational terms. Unlike the power as currency approach, it is not sufficient to measure the amount of power of each actor as indicated by his or her possession of selected variables, such as weapons, financial resources or personnel (Merritt and Zinnes, 1989). The relational definition of power also requires the measurement of the degree of need among other actors. To assess different degrees of power consistent criteria not only have to be developed for the evaluation of the degree of power provided by resource-dependence and institutions, but also for their combined effect. Contrary to the directional approach to power relations, the simple additive combination of the two dimensions is prohibited by such questions as whether resource-dependencies or institutions can overrule each other or whether and to what degree they enhance each other. The selection of consistent criteria for measuring and comparing degrees of resource-dependence and institutional power is obviously very problematic.

The British policy network approach has tended to circumvent the theoretical and practical difficulties of measuring different degrees of power in favour of subjective-descriptive evaluations of power relations (Rhodes, 1986). Thus, some inductive network models have resorted to questioning the members of the networks about perceived differences in their power. Other deductive network analyses have been based on the subjective assessment of power by academics. While these approaches allow for a more differentiated depiction of power relations in networks, they have been one of the main obstacles for the development of a network theory due to the problem of arriving at inter-subjectively agreed criteria. Conversely, network analysis has been hampered by a profusion of typologies of networks each based on different and rather vague criteria. Moreover, the preoccupation with the power structure of networks has led to the under-theorization of the concepts of agency and process in British policy network analysis as has been criticized by both the advocates of the approach and its critics.

The definition of power relations in terms of their causal direction not only avoids these problems, it also returns to the origins of network analysis which focussed on the position of actors within the structure of their network. While a number of sociologists have continued to develop this approach with highly theoretical models, political science has proceeded towards greater descriptive detail in the analysis of the individual relations in networks. As an example of the former, Karen S. Cook, Richard M. Emerson, Mary R. Gillmore and Toshio Yamagishi (1983, 1988) have examined how the simple presence or absence of relations among network actors bestows power upon those actors who have a high number of linkages and who are centrally placed within their network (Skvoretz and Willer, 1993). Conversely, the latter is represented by policy network models which emulate pluralist or bureaucratic decision-making models in seeking to explain the influence of network actors by descriptive accounts of the variegated characteristics of their relations (Waarden, 1992). Choosing a simple fourfold typology of causally directed network relations, multilevel network theory returns to the origins of network analysis. It redirects the focus of network analysis on how the position of actors within a network affects their interactions and their ability to influence the decision-making process (Dowding, 1995: 152).

Actors in Multilevel Networks

The preceding analysis of the power relations in networks takes the central place in network analysis because it is presumed that the actors will use their power in order to influence the decision-making process in their favour (Laumann et al., 1991: 63). The power structure of the network determines the ability of different actors to change each other's preferences regarding particular policies. However, in order to explain the resulting decision-making process, multilevel network theory has to make theoretically guided assumptions about the ways in which actors use their power relations within the network to exert pressure and when actors modify their policy preferences in response to pressure from other actors. In short, multilevel network theory has to illustrate the relationship between network structures and behaviour of political actors in the decision-making process.

The following suggests that three connected variables determine the decision-making process: the actors, their preferences and the calculations which guide their actions. In existing network models various ways have been proposed to conceptualize them. The following section examines which is best suited for analysing multilevel foreign policy decision-making. Moreover, this section proposes that multilevel network theory can fruitfully draw on rational choice theory to arrive at general hypotheses regarding the behaviour of political actors in the national as well as international domain.

The concept of actors in networks is crucial for an analysis of decision-making processes in two respects. First, the conceptualization of network actors determines their resources and needs in the analysis of a network's structure. Second, the concept of network actors influences our understanding of their political preferences and their behaviour in the decision-making process. Three competing concepts of actors can be distinguished in various network models: collective actors, individual agents and role actors. Each poses different problems for the analysis of multilevel decision-making.

Collective Actors

The concept of collective actors presumes that decision-making networks consist of relations among collective organizations, such as parties, interest groups, large firms, unions or governmental agencies (Atkinson and Coleman, 1989; Rhodes and Marsh, 1992: 9; Mayntz, 1988: 192). The advantage of this approach is that the assumptions and conclusions regarding the power and interests of collective actors can be generalized. Although the membership of collective actors is subject to constant or periodical changes, the power relations and functions which refer to the organizations rather than their individual members are relatively stable. Moreover, institutionalized collective actors hold resources independent from the contributions of their membership. These resources can be employed for purposes which lie outside the immediate interests of their members. Among these interests, the most important goal is that of organizational welfare. The interest of organizations in their continued existence regardless of the necessity to fulfil certain functions within the political, social or economic system can be explained by the division between membership and consumers on one hand and leadership and employees on the other. Since the human agents who are employed by an organization have a stake in its maintenance, i.e. their personal welfare, organizational survival is not merely an intermediate goal, but a primary objective in itself. Thus, collective actors do not only hold a stable position within a network as determined by their resources and institutional attributes, but also have a range of stable needs which derive from the independent and often prevalent goal of organizational survival and welfare.

For the development of a multilevel network theory of European foreign policy decision-making, however, the utility of the collective actor approach is limited. Specifically, it poses a problem for the consistent conceptualization of diverse actors at the national and international level within a single theoretical framework. Most crucially, the membership of collective actors at different levels of analysis varies between individuals, organizations and even states. As a consequence it is not possible to provide a consistent definition of collective actors for multilevel analysis. Additional disadvantages of the approach concern our definition of power which requires actors to use their relations with other actors intentionally. Rational choice theory has shown that collective actors fail to meet the requirement of intentional action since, as non-unitary actors, they do not necessarily have consistent preference hierarchies (Green and Shapiro, 1994: 15; Buchanan and Tullock, 1962: 13).

Individual Actors

An alternative to the collective actor concept is presented by the individualist approach. The individual actor concept models networks as linkages and communication lines between individual human agents (Heclo, 1978; Richardson and Jordan, 1979; Bressers et al., 1994; Wilks and Wright, 1987). In fact, any empirical study of decision-making in networks will deal with the interactions of individuals, not impersonal organizations. The individualist perspective avoids the problem of intentional action since goal directed behaviour is a distinctive feature of human agency. In addition, the individualist perspective recognizes that the ability to wield power also depends on personal capacities and characteristics. Similarly, the preferences of an actor may be determined by their organizational environment, but also by their personal desires. The descriptive and explanatory capacity of an individualist perspective, thus, is much higher than that of the collective actor approach.

By introducing additional variables related to the individual character of actors in decision-making networks, the analysis of specific cases is more comprehensive. Nevertheless, for the construction of a multilevel network theory the individual actor concept can only be rejected because it eventually inhibits theoretical generalization. If personal characteristics play a dominant role in defining the relations among network actors and their interactions, no general statements can be made about them. The usage of the individual actor concept by network models has, therefore, contributed to its limitation to empirical-descriptive analysis which has been criticized in the introduction of this book.

Role Actors

The preceding argument suggests that in the conceptualization of network actors a choice has to be made between theoretical generalization and descriptive detail. Although this is ultimately true for the empirical analysis, both can be accommodated within the framework of multilevel network theory. The two perspectives can be reconciled by the concept of actors as individuals who play political, social, or economic roles (Dowding and King, 1995: 11; Waarden, 1992: 33). A role is defined by the rights and obligations attributed to it by formal and informal institutions, its command over resources and the expectations which the role player and other members of a system or organization hold with regard to it. Roles can only be understood in their institutional and social context. However, roles are held by individuals. They provide a conceptual bridge between the individual who bears a role and the social collective which shapes it. Moreover, the concept of roles can be employed in general theoretical accounts as well as in detailed empirical analyses.

On the level of a general network theory, roles are sufficiently defined by their enduring features. These are the aspects by which any individual who impersonates a role will be affected. Ranking in order of stability, formal institutions prescribe the most enduring attributes of roles. Informal institutions, resources, perceptions and expectations complement them. In all these respects, roles are crucially related to collective organizations and hence to the collective actor approach. In modern societies, organizations prescribe the institutional setting for roles, provide resources to enable them to fulfil their functions and shape the expectations regarding roles as captured in the notion of organizational cultures (Dowding, 1991: 147; Searing, 1991: 1245). These variables not only define general features of actors in decision-making networks, empirical studies also suggest that they dominate individual preferences and behaviour. Thus, it can be argued that individuals have to fulfil their role in order to obtain personal objectives (Searing, 1991: 1254). In fact, individuals are commonly appointed to a certain position because they meet the cultural expectations connected with it. For instance, bureaucrats, politicians as well as managers often share not only similar norms, cultures and preferences, but also aspects in their personal history, such as a university education. Moreover, roles shape the behaviour and preferences of human agents through internalization as they learn to meet the standards and expectations which constantly confront them (Bressers et al., 1994: 6). In sum, the concept of actors as role players can accommodate both theoretical generality and descriptive accuracy.

Interests and Preferences

The key question which multilevel network theory seeks to answer is how actors are able to influence each other's policy preferences and ultimately the outcome of the decision-making process. In order to do so, it has to distinguish between the political preferences of actors in the absence of external influence and those which result from changes due to intentional pressure from other members of the network. The former is usually referred to as interests, while the latter will here be termed preferences (Dowding, 1991: 30). While preferences and their changes can only by examined empirically, the interests of network actors can be established either inductively or deductively. The following argues that the inductive approach is preferable because of the problems associated with the identification of objective interests.

The deductive analysis of interests is commonly based on an examination of the needs of actors. It presumes that the policies which enable actors to ensure their welfare and fulfil their functions define their objective interests. While such assumptions underlie a number of theories in international relations - in particular models which treat states as unitary actors - it conflicts with the basic concepts of multilevel network theory. Specifically, it encounters the problem of multiple roles mentioned above. Different roles often have conflicting objective interests. Since individual human agents typically hold multiple roles, the deductive determination of interests would require additional hypotheses about the way in which agents resolve the conflicting interests of their diverse roles. More critical in terms of network analysis is the problem which arises from the notion of mistaken interests (Barry, 1965/1990: 179). The deduction of interests from needs allows actors to be mistaken about their objective interests, i.e. if they do not recognize their need for specified tangible and intangible commodities (Dowding, 1991: 36). This concept fundamentally contradicts network analysis which requires that actors are aware of their interests. Only if actors are, can they influence the decision-making process *intentionally* as required by the definition of power in the previous section.

As a consequence, it can be contended that an inductive approach to the analysis of interests is more appropriate for multilevel network theory. According to this method the interests of actors can be inferred from their publicly expressed preferences. This approach avoids the concepts of original or mistaken interests. It proceeds from the premise that interests can only be truly known to individuals themselves. The analyst has to contend with whatever preferences actors chose to make public. Obviously, these preferences may change. Moreover, actors might adjust their public preferences in order to pre-empt a controversy (Connolly, 1983: 49). Since multilevel network theory seeks to establish the effects of intentional influence on the decision-making process, however, it is not relevant whether the observed preferences at the beginning of a research period are the original interests of political actors. Multilevel network theory should be able to explain the actions and preference changes at any possible (starting) point of the decision-making process. While it would be generally desirable to trace the decision-making process from its perceived beginning as marked by the emergence of a particular issue or problem perhaps, the hypotheses of multilevel network theory which concern the actions and preference changes of network actors should be valid at any stage of the political process. The basis for these hypotheses is rational choice.

Rational Action in Multilevel Networks

It has been stated above that multilevel network theory proceeds from the premise that political actors seek to ensure that their political preferences will be served by the outcome of the decision-making process. In order to do so, actors attempt to influence each other and, finally, the ultimate decision maker. The interactions which evolve due to these attempts are a result of the structure of the network on one hand, and the distribution of preferences with regard to a political issue on the other. However, neither does pressure always lead to preference changes, nor do network structures prescribe a single course of action in order to influence the decision-making process. Actors can choose among their network linkages. Most crucially, actors choose whether to change their preferences and join a coalition in favour of a particular policy. These choices are strategic choices since they depend on the choices and behaviour of other actors within the network. By hypothesizing about the choices of network actors, multilevel network theory proposes a causal link between the structure of the network and the behaviour of political agents in the decision-making process. The following examines the axioms on which hypothesis regarding the choices of preferences and actions can be based. It suggests that cost-utility calculations derived from rational choice theory can provide a range of hypotheses which illustrate how actors may utilize their position in multilevel networks in order to influence the decision-making process.

Rational Choice

Rational choice theory posits that human agents can be modelled as calculating actors who pursue cost-utility optimizing behaviour (Dunleavy, 1991: 3; Green and Shapiro, 1994: 14). That is actors choose rationally if they select the behaviour or preferences which they believe will yield their desired outcome at the lowest cost (Hollis and Smith, 1991: 272). The problem of analysing the expectations of different actors regarding the rationality of different options 'under due consideration of the circumstances' (Oppenheim, 1981: 126) has led to the introduction of the concept of bounded rationality (Elster, 1986: 5). Actors assess the costs of only those options of which they are immediately aware since the investigation of all possible alternatives is too costly. As such, the concept of bounded rationality is closely related to network analysis. By definition networks are described as stable and routinely used linkages among a set of actors. As a consequence, the members of a network have a clear understanding of the nature of their power relations with others and whether they have the potential to influence other actors.

The value of the rational choice approach for network analysis has been pointed out in various studies (Benz, 1993; Dowding, 1995; Daugbjerg and Marsh, 1998). It lies in the fact that rational choice proposes a general principle for choosing strategies and preferences. It fulfils on one hand the requirement of generalization which is the basis for a theory of choice, on the other it is characterized by variability and specificity. Moreover, if rational choice is defined in terms of cost-utility calculations, it enters any theory which acknowledges that actors choose among various options for action by considering the required resources and the likelihood of succeeding to obtain their objective. Many international relations theories and models of decision-making implicitly or explicitly refer to such calculations to explain the behaviour of political actors. One reason for its wide usage seems to be that rational choice is inherent in the concept of power as it has been defined at the beginning of this chapter. The assumption that power can be deduced from bases such as resource-dependencies and institutional relations is grounded on the notion that A can and presumably will impose costs on B, if B does not comply with his or her wishes (Connolly, 1983: 102; Harsanyi, 1962). An analysis of decision-making networks in terms of power relations, therefore, suggests a theory of agency based on rational choice.

Cost and Utility

In order to employ the concept of rational choice to explain the behaviour of actors in multilevel decision-making processes, network theory has to specify two factors: the utility of different behavioural strategies to an actor and their relative cost. In a network of power relations, the utility of a network linkage is defined by the location of the ultimate decision unit, i.e. the role actor or actors who have the formal institutional authority to make legitimate and binding political decisions regarding a particular issue. The further the ultimate decision unit is removed from an actor, that is the more actors serve as intermediates, the weaker is his or her power and the smaller the utility of pressure exerted through these relations. Although the network position of the actors themselves is stable, the utility of an actor's linkages can vary because of changes in the ultimate decision unit. The location of the ultimate decision unit not only changes from issue to issue, but can also shift as a result of the interactions among network actors during the decision-making process. Typical ultimate decision units are ministers for routine issues, cabinets and parliaments for important or controversial decisions and international organizations for multilateral actions. Most issues start at lower levels such as ministries, but some might move up to the cabinet or even international organizations because of internal dissent or the inability to provide adequate solutions at a sub-national or national level.

The cost of an action is determined by the type of power relationship with each actor. Several premises regarding the relative costs of each type of power relation can be derived from rational choice theory. First, the exercise of pressure is always costly (Keohane, 1984: 89). Not only do actors have to invest in communication, they also have to consider the costs of using their resources or institutional authority in order to exert pressure on another actor. From this follows the basic premise that actors will only seek to influence the decision-making process if their preferences are affected and the cost of an adverse policy outcome is higher than that of interaction (Mayntz, 1988:189; Elzen et al., 1990: 186; Wright, 1988: 596). The costly initiative lies, therefore, with the actors who perceive their policy preference to be threatened or in a minority. They have to engage in the decision-making process in order to increase the support for their preferred policy outcome. Conversely, actors who are part of the majority view will refrain from pressing other actors to support them until their preferred policy outcome is seriously threatened. As a result, the number of network actors engaged in the decision-making process should rise over time as more and more actors are pressed to take a stance for or against a policy (Mayntz, 1988: 207; Atkinson and Coleman, 1992: 160). Thus, the first premise states: Actors will only seek to influence the policy preferences of other actors in the network if they perceive their preferred policy to be in a minority.

Second, A's costs for exercising pressure are lower if A has power over B or is interdependent with him or her, than if A is subject to B's power or the two actors are autonomous. Indeed, the exercise of pressure through relations which give B power over A is prohibitive for A because he or she cannot inflict any cost on B to support it. Autonomous relations prevent the exertion of influence due to the absence of an established institutional or resource exchange relation. Although a coalition cannot be ruled out in the case of similar interests, the cost of establishing a new relationship can be regarded as higher than the usage of already existing network linkages. The costs of each power relation can be summarized in form of a simple hierarchy: Low costs for A are associated with relations in which A has power over B [A>B] or A and B are interdependent [A<>B], while high costs are linked to relations were B has power over A [B>A] or where both actors are autonomous [A|B]. Accordingly, the second premise suggests: Actors whose policy preference is in a minority will use their power over dependent actors B [A>B] or their interdependent relations [A<>B] to exert pressure on actors who hold different policy preferences or who are undecided.

Finally, choice theories at all levels of analysis have recognized that legitimized pressure is less costly than not legitimized (Keohane, 1989: 5; Young, 1980: 338). That is, actors who have recognized institutional authority over another's actions have lower costs in trying to influence them than actors without legitimate control. For instance, ministers have lower costs influencing their staff than representatives of interest groups. The difference is of particular interest in collective decision-making units, i.e. collective political bodies with institutionalized legitimate decision-making rules, such as majority voting or consensus. In collective decision-making units, most of which are national parliaments or the ministerial councils of international organizations, the formal institutional influence of members can thus be differentiated from the informal, and non-legitimate, influence of actors who are merely linked to the collective decision-making unit. Members with a voice or veto in a collective decision-making unit have lower costs in influencing decisions within the body than nonmembers. Hence the final premise assumes: The legitimate pressure of members with a simple majority or veto position in a collective decision-making unit is less costly than the pressure from actors who are only linked to the organization.

Acting Rationally

The above premises postulate how actors use their power relations within the network in order to influence each other and the decision-making process according to Table 2.2.

Table 2.2   Rational action

	A has power over B [A>B]	B has power over A [A<B]	A and B are interdependent [A<>B]	A and B are autonomous [A|B]
A and B have similar preferences	A and B form coalition	A and B form coalition	A and B form coalition	no interaction
A and B have different preferences	A exerts pressure on B	B exerts pressure on A	A and B exert pressure on another	no interaction

Based on these assumptions multilevel network theory proposes two hypotheses which specify when actors are likely to succeed in changing each other's policy preferences and influencing the policy outcome. From the simple additive cumulation of the potential cost which can be imposed upon a network actor by those who have power over him follows the first hypothesis:

> The higher the degree of pressure (P, in per cent), i.e. number of actors exerting pressure (E) on a single actor X out of all actors who have power over him or her (L), the more likely is actor X to change his or her policy preference.

According to the first hypothesis actors who are exposed to higher pressure to change their policy preferences are more likely to modify their position than those who are subject to lower degrees of pressure. The degree of pressure P on actor X during a phase T, which is delineated by the timing of two preference changes T-1 and T in the network, is calculated in the form of:

$$P_T [X] = E / L$$

To take an illustrative example from the case of air strikes in Bosnia, seven out of 39 actors who were directly linked to the British Prime Minister [PM] pressed him in May 1992 to support air strikes in order to contain the Serb advances in Bosnia. According to the formula, P_1 [PM] = 7/39 = 18%, this amounted to a degree of pressure of 18 per cent. By comparison, 28 per cent of the actors who are able to exert power over the American President [Pre] urged him to adopt air strikes in May, namely nine out of 32 with P_1 [Pre] = 9/32 = 28%. Following the proposition of the first hypothesis, the American President was, therefore, more likely to change his preference in favour of air strikes than the British Prime Minister. Indeed, as the case study will show, President Bush publicly endorsed air strikes in the following month, while Prime Minister Major resisted the calls for military strikes until spring 1993 by which time the pressure on him had increased to 36 per cent.

All actors (L) who have power over another actor X are also called his or her constituency. It is important to note that each actor has a different constituency, since each is linked to different actors in the network. As a consequence, the degree of pressure in favour of a particular policy is always relative with regard to the target of the pressure. Some actors are subject to the power of fewer actors in the network than others and can thus be described as relatively insulated from external pressure. However, this often means that they also have fewer linkages through which they themselves can exert pressure for their preferred policy. Other actors, in particular actors with multiple or boundary roles, are more tightly integrated into the network by means of a large range of linkages. These tend to be exposed to more pressure, but are on the other hand frequently able to use their relations to mobilize pressure.

Crucially for the explanation of the decision-making process, the role actor who forms the ultimate decision unit for a specific issue or at a certain point during the decision-making process is subject to the same behavioural rules as other network actors. Since the policy outcome is the preference of the ultimate decision unit, it is determined by the degree of pressure to which this actor is exposed from other actors in the network. The concept of the winning coalition, which has been used in network models as well as other multilevel theories, here always refers to actors who are directly linked to an ultimate decision unit, i.e. his or her constituency.

Actors who are not directly connected to a decision-maker cannot influence the outcome, except indirectly through a series of preference changes which involves actors who are directly linked to the ultimate decision unit. Since most actors are not directly linked to the ultimate decision unit, the decision-making process becomes therefore an essential element in the explanation of the formation of a winning coalition and the policy outcome.

However, the first hypothesis can be qualified according to the third premise to form the second hypothesis:

> Collective decision-making units can resist higher degrees of pressure than role actors if members use a veto or if a decision requires a (qualified) majority.

The second hypothesis, which will be referred to as the veto or blocking strategy in the following case studies, suggests that collective decision-making units, such as parliaments or international organizations, will typically modify their policy preferences at higher degrees of pressure than unitary role actors if one or several members use a veto or if a required majority blocks a preference change. In order to explain this difference in behaviour, collective decision-making units are best understood as a network within the network. Internally, the decision-making process within collective decision-making units is defined by the first and second hypothesis. However, externally collective decision-making units act as a unitary role actor within the network. In order to do so these organizations have to reach a certain degree of consensus in order to arrive at a single policy preference which is then supported and expressed by the collective body mainly in the form of communiqués.

In addition to increasing the ability of collective decision-making units to resist network pressure, the blocking of a decision-making unit influences the decision-making process in that it can cause the issue to be referred to another decision unit, usually with higher institutional authority. For instance, the second case study will show that the inability of most European organizations such as the EU and the WEU to agree on a decision to intervene in Bosnia led to the transfer of the responsibility over the international response to the Yugoslav crisis to the UN Security Council in summer 1992. In fact, the case reveals that the French government intentionally blocked a decision in these organizations because it wanted the UN Security Council to hold the ultimate authority over an international intervention in the former Yugoslavia. Its reasons were twofold. First, France had, as one of the five permanent members of the Security Council, greater influence on its decisions than other actors. Second, the balance of preferences within the Security Council was less in favour of air strikes than within the EU and the WEU.

Operationalization

Based on the above discussion of the nature of power in networks, the map of the multilevel European foreign policy network used in the following involved all national, transnational and international actors who were regularly engaged in the European foreign policy decision-making process between 1990 and 1995. Broadly speaking, the European foreign policy network included all domestic governmental, administrative and party political role actors, such as the heads of state, foreign, defence and economics ministers and their departments, the main political parties within the national parliaments, the representatives of employers associations, unions, and interest groups in Europe and North America. In addition, it covered representatives and officials of the European Commission, the European Council of Ministers, the key party political groups of the European Parliament, the North Atlantic Council and NATO's integrated organizational body, the WEU, the OSCE and the UN Security Council and Organization. Notably, the US and Canada were part of the network through bilateral relations with core European states and their membership in the preceding international organizations. For each of these role actors the institutional and resource-dependence relations with any of the other actors was examined and mapped in terms of the four types of power relation - A>B, A<B, A<>B and A|B - in a two-dimensional matrix which can be found in the appendix.

The subsequent analysis of the interactions and policy preferences changes of national, transnational and international actors in the European foreign policy decision-making process was based on primary and secondary documentation of the cases. In particular, the data included policy drafts, parliamentary debates, party programmes, press statements and organisational declarations. In addition, the analysis covered 25 British newspapers, Jane's Defence Weekly, the *Frankfurter Allgemeine Zeitung* and *Reuters German News* as well as the newspaper clip collection of the Freie Universität Berlin which provided relevant articles from the *Frankfurter Allgemeine Zeitung, Frankfurter Rundschau, Neue Züricher Zeitung, Das Parlament, Süddeutsche*

Zeitung, *Die Welt* and *Der Spiegel*. Specifically, the three case studies distinguished four types of behaviour among the members of the multilevel European foreign policy network: no change of preferences, unclear or undecided preferences, change of preferences and the veto or blocking of a preference change in collective decision-making bodies.

3 European Union: The Dual-Use Control Agreement

Introduction

A broad number of national, transnational and international actors contribute to the making of European foreign policy today. The European Union is not the only one. However, the EU plays an increasing role in defining multilateral and unilateral foreign policies in Europe.

This chapter examines how multilevel network theory can help to understand the complex interactions between the EU and its member states and their impact upon European foreign policies at the national and international level. As an example, this chapter analyses the making of a common European framework for export controls concerning goods with civil and military applications ('dual-use') and specifically the German resistance to a reduction of its national controls as a consequence. Moreover, it observes how the international context, notably the support of the US for European dual-use export controls and the redefinition of multilateral export control regimes after the end of the Cold War contributed to defining the options for a common European foreign policy on dual-use transfers.

The issue of common dual-use export controls within the European Union emerged in the early 1990s due to the impending implementation of the internal market which threatened to undermine existing national controls. Incidentally, it followed a series of export scandals in the late 1980s which had led to a tightening of dual-use export controls of many European countries, such as Britain, France, Italy and the Netherlands. In particular, Germany had increased its dual-use regulations after news emerged in 1989 that German companies had played a significant role in the construction of a chemical weapons factory in Rabta, Libya. However, while initially stronger national and international controls of sensitive exports were widely supported in Germany and elsewhere in Europe, by 1992 industry pressure began to mount for a relaxation of the regulations. The German industry specifically sought to use the negotiations over common European dual-use controls after the implementation of the internal market to achieve a reduction of Germany's tight national regulations. Crucially, for the industry the governments of Britain and France insisted that any common European dual-use transfer controls would have to be based on the lowest common denominator among the member states. Over two years the German government resisted these pressures. However, by the end of 1994, the German government unexpectedly not only acceded to the common framework for European dual-use transfers, but in fact reduced its controls beyond the requirements of the common regulations.

Important for the testing of multilevel network theory, exogenous factors cannot explain the German foreign policy change. The danger of conventional, nuclear, chemical and biological proliferation did not decrease so as to justify less restrictive dual-use export controls. While the end of the Cold War reduced perceptions of threat from the former members of the Warsaw Treaty Organization, the dissolution of the Soviet Union and the sale of its military technology to Third World countries actually increased the dangers of international proliferation. Moreover, the Gulf War and the conflict in the former Yugoslavia illustrated that a multipolar international system was not necessarily more peaceful. Rogue states in the Middle East or Eastern Europe showed that they were willing to use the weapons they had acquired to achieve political and military aims. Furthermore, although the governments of Germany and other Western industrialized countries enhanced their export controls, the series of arms export scandals continued unabated after the Rabta affair. In particular, the UN investigation into the Iraqi weapons programme after the end of the Gulf War provided a continuous stream of information about the contribution of American and European companies to the build-up of Third World weapons arsenals.

An analysis of the making of German foreign policy in this case is especially interesting because the argument put forward by the German administration to justify its control cuts; namely that the control reductions were required to conform with the new common European dual-use regulations, was insufficient, if not incorrect. The German Cabinet not only decreased its controls before, but also beyond the European agreement on dual-use controls on 19 December 1994. Notably, an opt-out clause had been agreed which allowed member states to maintain stricter national export regulations in some areas if they were regarded as necessary to safeguard national security interests.

Specifically, the German government could have preserved its export restrictions for certain volatile countries which were identified in the so called 'H-list'. The H-list was the most efficient means of dual-use export controls and had been introduced in Germany only in 1992. The Kohl Cabinet's decision to drastically shorten Germany's country list even before the EC regulations came into place indicates that the common European controls were neither the only nor the main reason for the government's policy change. Moreover, the observation that the British government maintained its country list under the European agreement increases the doubts over the explanation offered by the German government.

This chapter is structured in four parts. The first part outlines the circumstances which led to the beginning of EC negotiations over common dual-use controls in the early 1990s. The second part provides a detailed analysis of the foreign decision-making process in Germany and the EC between 1992 and 1995 with specific reference to the preference changes of key actors and the degrees of pressure to which they were exposed at the time. The third part offers a quantitative assessment of the data with regard to the hypotheses proposed by multilevel network theory, and the final part examines which new insights into the making of European foreign policy could be derived from this case study.

Dual-Use Controls in the Early 1990s

The reconsideration of European dual-use export regulations is best understood in the context of the national and international conditions which defined the agenda and the options with regard to the development of comprehensive and restrictive national dual-use controls in Europe between 1990 and 1992. The initial tightening of dual-use export regulations in Germany followed the Rabta affair at the beginning of 1989. Intelligence information published by US American news sources revealed that German companies had sold dual-use technology to Libya where it had been used to build a chemical weapons factory in Rabta. The reaction to the news was outrage, both abroad and in Germany (Wulf, 1991: 83; Hantke, 1992: 257). The heated international and domestic debate about the need for stricter export controls for dual-use equipment which followed was fuelled by the discovery that German firms had also contributed to the military build-up of Iraq. After the Iraqi invasion of Kuwait in 1990-91, the fact that German soldiers did not participate in the Gulf War alongside contingents from the US, France and Britain contributed to the international criticism of Germany's foreign policy. In reaction to national and international demands for tighter controls on dual-use equipment, the government under Chancellor Helmut Kohl introduced a range of amendments to the Weapons of War Act and the Foreign Trade Act between 1990 and 1992. Among others, the German government introduced the 'H' country list which established special controls for dual-use exports to 54 states (Bauer and Eavis, 1992: 7; Wulf, 1991: 75). Although the list was shortened to 35 after the end of the Cold War, the new German export regulations for dual-use equipment were now among the strictest in Europe (Bauer and Eavis, 1992: 7).

However, German companies had not be alone in exporting extensively to sensitive regions in the Middle East. By 1992, the British administration was compromised by the Supergun and Matrix Churchill affairs and the US Commerce Department was accused of having ignored warnings from the Pentagon over American arms sales to Iraq and Jordan. As most European governments and the US were forced to acknowledge that they were by far free from blame, the tightening of export legislation in Germany was matched by similar developments in Britain, France, Italy, the Netherlands, Belgium, Denmark and the US.

The range of international anti-proliferation agreements was also expanded. Alerted by the possibility of constructing a chemical weapons factory from fertilizer plant components, the Australia Group, an informal group of 31 countries committed to chemical and biological non-proliferation, extended its controls to cover dual-use equipment and technology which could be used to manufacture chemical weapons (Lundbo, 1997: 138). The Australia Group, which included among others Germany, Britain, France and the US, also expanded its list of precursor chemical substances from nine to fifty (Handke, 1992: 260). The discovery of the Supergun project in Iraq triggered a revision of the Missile Technology Control Regime among a similar set of member states. The regime had been established in 1987 and issued common guidelines for the export of missiles and related equipment, material and technology. The transfer of nuclear related technology was in turn further regulated by the Nuclear Suppliers Group and the Zangger Committee, two informal groups of countries which, unlike the Australia Group, not only included most European states and the US, but also Russia (Lundbo, 1997: 141; Bauer and Eavis, 1992: 14). At their meeting in Warsaw on 3 April 1992, the Nuclear Suppliers Group specifically agreed on common export regulations for dual-use goods with nuclear applications.

In spite of the broad consensus underlying the unilateral and multilateral tightening of dual-use export regulations between 1990 and 1992, the impending implementation of the European single market on 1 January 1993 put the new export controls almost immediately into question. The dismantling of border controls and licensing procedures for the European Community (EC)[1] internal transfers threatened to undermine not only national export regulations on armaments and dual-use technology all over Europe, but also the multilateral regimes which were all nationally implemented. If effective controls were to be maintained, EC member states had to coordinate their policies. How this was to be achieved was a contentious issue among the member states. While EC members agreed that weapons exports should remain strictly under national authority according to article 223 of the Treaty of Rome, the distribution of competences was not so clear in the case of dual-use goods because they had primarily civilian applications.

The EC member states had three options. First, they could treat the transfer of dual-use equipment as weapons exports and continue their national controls. Second, EC members could dismantle their national regulations and replace them by common European controls which would apply to all exports outside the EC. Finally, member states could seek to establish a multilateral control regime for dual-use equipment which included not only the EC, but also the United States and other industrialized democracies. The impending renegotiation of the COCOM regime which had regulated dual-use exports of Western industrialized democracies during the Cold War presented an opportunity for the latter. During the first stage of the decision-making process, all three options were still under consideration. However, as the following will show, the pressure from various national and international actors in the European foreign policy network soon narrowed the debate down to the establishment of common EC regulations.

The Dual-Use Control Decision

The beginning of the debate over a revision of European dual-use controls in view of the internal market immediately followed the approval of the German Bundestag for the nineteenth amendment of the Foreign Trade Act on 14 February 1992 which completed the tightening of the German dual-use export control system. While the distribution of preferences had been very much in favour of restrictive controls during the previous two years, the negotiations in the EC over common regulations for dual-use exports considerably changed the conditions which had led to the establishment of more restrictive German legislation and other EC member states. Most importantly, it gave EC member states, which on the whole had maintained weaker controls than Germany, a direct interest in the level of the German dual-use export regulations.

Before the start of the European negotiations the German industry, represented by its national associations such as the German Chambers of Commerce and Industry (DIHT) and the Federation of German Industries (BDI), had been isolated in its advocacy of less restrictive export rules. In fact, key figures of the German industry themselves had initially supported stricter export controls after the scandals in Iraq and Libya. The international outrage at media reports of German armaments exports had been perceived as detrimental to the industry's reputation. However, as early as 1991, industry spokespersons had criticized the extent of the new regulations because it reduced the industry's international competitiveness.

The passing of the nineteenth amendment of the Foreign Trade Act, however, showed the resolve of the government and parliamentarians from all parties that German companies should not again be allowed to become involved in the military build-up in the Middle East or elsewhere in the world. Successive export scandals kept the public wary about the assurances of the industry that there had merely been a few 'black sheep'. Due to the widespread national consensus regarding the tightening of the controls, the industry had not been able to prevent or change the revision of the German export regulations. With the start of negotiations about common European dual-use export controls, the industry instantly received support for controls at the level of the lowest common denominator from most other European countries where export regulations were less restrictive than in Germany.

The fact that defence-related exports were a national prerogative according to article 223 of the EEC Treaty complicated the negotiations. It meant that a decision could not be taken within the regular decision-making framework of the EC, but had to be agreed upon in intergovernmental bargaining. Crucially for an understanding of the case, the intergovernmental nature of the decision-making process provided all EC governments with the ability to veto or block the outcome of the negotiations. This enabled the EC governments, including the German administration, to resist higher degrees of pressure than could have been expected under other circumstances. Moreover, the veto position of the German government had direct impact on the strategies which the advocates of less restrictive dual-use transfer laws could rationally pursue. Specifically, the German industry associations could not expect that European resistance to strict common dual-use regulations would force the German government to accept lower national standards. The Kohl government could veto or opt out of the European scheme at any time if it believed the controls to be insufficient. Moreover, the German government retained the ultimate decision-making authority over its national export controls. In order to reduce the German controls the industry had to increase the active support for a revision of the Foreign Trade Act in Germany.

German Industry Pursues Twofold Strategy

In accord with these considerations, the German industry pursued a twofold strategy. On one hand industry representatives used their transnational linkages with European Commission and European Parliament members to lobby for common European dual-use transfer regulations. The negotiations would ensure continuous pressure of Germany's European partners on the Kohl administration. An example of this strategy was a letter by the chairman of the Daimler-Benz AG Edzard Reuter addressed to Chancellor Helmut Kohl and the

President of the European Commission Jacques Delors which called for joint European export controls on dual-use equipment as early as 1991. On the other hand, representatives of the BDI, the DIHT, sectoral industry associations and the largest German technological companies mobilized their relations with other domestic actors such as the members of the Christian Democratic Union (CDU), its Bavarian sister party the Christian Social Union (CSU) and the liberal Free Democratic Party (FDP), officials in the Economic, Foreign and Defence Ministries and cabinet ministers to argue that Germany should conform with the lower dual-use export control standards of its European neighbours.

The influence of industry representatives was based on the crucial role of manufacturing for the German economy. Particularly sensitive were regions such as Lower Saxony and Mecklenburg-West Pomerania where disproportionately large sections of the electorate were employed in technical and manufacturing industries. Given that unemployment had increased considerably due to German reunification and the problems associated with the economic restructuring of the former East German Länder, the threat of further redundancies because of decreasing exports was very persuasive. In this line, spokespersons of the DIHT complained that the German dual-use controls deterred potential costumers because they were not sure to receive export licenses. The Bavarian Economic Minister Georg von Aldenfels (CSU) agreed. However, the role of the Länder governments in the decision-making process was limited. Since export laws were exclusively under the authority of the federal government, the Länder governments could at best seek to influence the administration, and thus the decision, indirectly. Economics Minister Jürgen Möllemann (FDP), however, rebuffed the complaints from industry representatives and some Länder ministers. In the light of the recent experiences with illegal armaments exports the government perceived the legislation as necessary in the interest of national and international security.

EC Commission Proposes External Fence

While German companies were the only domestic actors in the European foreign policy network who unreservedly supported a reduction of the German dual-use controls in February 1992, the beginning of the intra-European negotiations soon led to additional international pressure. It emerged quickly that most EC governments rejected common European dual-use controls at a level comparable to the German standard. In the EC, the German government's demands for comprehensive dual-use export regulations were only supported by the Italian administration. However, the members of the European Commission and a majority in the European Parliament also favoured high standards for dual-use export controls. The fact that the Commissioner in charge of the internal market was the German Martin Bangemann almost certainly contributed to this position on dual-use exports.

The influence of the European Commissioner on the intergovernmental negotiations was initially strong because the Council of Ministers decided to task Bangemann to produce a first draft for a common dual-use export control system in January 1992 (Bauer and Eavis, 1992: 11). Bangemann's proposal was accepted by the Commission later that month and communicated to the Council of Ministers and the European Parliament. The draft proceeded from the premise that dual-use equipment was not defence technology under article 223 of the Treaty of Rome and that intra-EC restrictions on the free movement of dual-use goods should end with the implementation of the internal market. In order to prevent the export of technology with civil and military applications to sensitive destinations, Bangemann suggested the creation of an external fence of controls for transfers outside the EC. The controls would apply to certain technologies and countries which were to be agreed by all member states.

Member States Disagree over First Draft

The German government welcomed the proposal which matched the German Foreign Trade Act. Moreover, the German administration agreed in principle with the Commission that the authority over common dual-use export regulations should lie with the EC. Nevertheless, German representatives warned that their political leadership would not compromise on the content of the lists. In the first discussions of the draft, the disagreement between the German delegates and their European colleagues was more fundamental. The British and French representatives outrightly rejected the introduction of a common country list (Bauer and Eavis, 1992: 12). To settle this dispute and to negotiate the substantive contents of a common regulation, representatives agreed to pass on the issue to a high-level working group staffed by national officials (Bauer and Eavis, 1992: 11). The strategy of transferring direct authority over the drafting process from the Commission, which was in favour of a comprehensive and centralized dual-use export controls system, was designed to remove the issue from the influence of the Commissioners. Moreover, the intergovernmental working group further institutionalized the existing close network among the EC administrations and provided through regular meetings additional opportunities to co-ordinate the opposition against high controls standards in the form of an intergovernmental coalition. Led by the British and French delegates, this intergovernmental coalition was able to exert considerable pressure on the European Commission and the German administration to accept the lowest common denominator as the basis for the European regulations during the course of summer 1992 (Bauer and Eavis, 1992: 11; Cornish, 1995: 40).

The Commissioners who were directly selected by the national governments and consequently sensitive to their demands were under the highest pressure within the European foreign policy network at 82 per cent.[2] The degree of pressure helps to explain why the Commissioners were the first actors within the network to succumb to the international opposition to strict controls for dual-use exports. When the Commission submitted its second proposal on the basis of the deliberations in the intergovernmental working group, the framework for the common regulations had been significantly watered down. The new proposal accepted that the lists of sensitive goods, destinations and licensing criteria were to be decided unanimously by the member states. However, the direct pressure from the intergovernmental coalition in the working group also affected the ability of the German representatives to implement and maintain their preferences with regard to the common dual-use goods export regulations. The staff from the Economics Ministry, which led the German representation in the working group, was especially vulnerable because of their constant exposure to the pressure from their European colleagues who accounted for 39 per cent of their relations within the network.[3]

Germany's European partners were also able to exert additional indirect pressure on the government through their linkages with staff in the German Defence Ministry. Although the Defence staff were not immediately involved in the international negotiations, they had a central interest in the debate since the level of the controls would considerably impact on collaborative armaments projects within NATO. Moreover, the German Defence Ministry staff was a strategic target for pressure from fellow European military officials since they had fewer linkages within the network than their colleagues in the Economics Ministry. Thus, the direct pressure from European Defence Ministry staffs and the Germany industry amounted to 40 per cent of the German military's contacts in the network.[4] The pressure on the heads of the two departments, Minister of Economics Möllemann and Defence Minister Volker Rühe was significantly lower at 34 and 31 per cent respectively.[5]

Commission Proposal Weakened

Although the staff from the two ministries maintained their support for the introduction of common EC regulations similar to the German Foreign Trade Act during the summer of 1992, the German negotiators had to concede on a range of issues as a consequence of the high international pressure. In spite of the veto position of the German government, they were only able to extract one major concession: the new proposal included a catch-all clause similar to §5c of the German Foreign Trade Act. The catch-all clause subjected any technology which to the knowledge of the exporter was intended for military use to controls. However, the German negotiators were not able to secure the regulation of knowledge and service transfers in the proposal. In addition, they had to accept concessions on the content of the two lists regarding dual-use technology and country restrictions. Four groups of sensitive goods were excluded from the common European list. However, it was assured that these could remain under national controls. In spite of the new proposal, the German administration was forced to admit by September 1992 that due to diverging national preferences a compromise about the regulation of dual-use exports was not in sight. In particular, the representations of Britain and France were dissatisfied with the new draft. Nevertheless, the German government resisted pressure to further compromise on its policy during the autumn of 1992.

The new proposal was also criticized in the European Parliament. However, for contrary reasons. A substantial number of European Parliamentarians attacked the Commission proposal because it allowed the member states with the least restrictive trade controls to determine the common European standard. In order to assure that tight criteria for the controls were introduced, the French Socialist Gerard Fuchs suggested transferring the authority over the control of dual-use exports to the Commission. Fuchs had been tasked by the European Parliament's Committee on Economic and Monetary Affairs and Industrial Policy to report on the Commission draft. However, his report had little impact. While the members of the European Parliament could exert pressure over the Commission and the political parties within each member state, they lacked direct influence over the European governments at the negotiation table. Although a majority in the European Parliament adopted a range of amendments to the draft and presented these to the European Commission, its views on the issue were disregarded.

German Government Seeks to Strengthen International Controls

Given the predictable difficulties in the European negotiations over common dual-use export controls, the German government simultaneously pursued the tightening of multilateral dual-use export controls in the wider international community. A possible ally was the US government which was not only in favour of common European controls in order to simplify American exports to the EC, but also among the few countries who sought to strengthen international export controls. On 15 June 1992, the US government announced that it was tightening its controls on missile-related technologies to 21 countries in order to encourage the implementation of the Missile Technology Control Regime. The decision was followed by an agreement among the members of the regime at their 29 June-2 July meeting to extend its scope to missiles capable of delivering biological, chemical and nuclear weapons.

With regard to dual-use technology US pressure for controls was more selective. The US administration was primarily concerned about exports to Iran, Iraq, Libya and North Korea. Thus, in October 1992, senior officials from the US State, Defence and Commerce Departments made trips to Japan, Britain, France, Germany, Italy and the Netherlands to try to persuade these governments to ban dual-use sales to Iran. However, even with regard to specific countries such as Iran, there was little support for dual-use transfer restrictions in these countries. While nuclear proliferation and the export of dual-use goods with nuclear applications were high on the political agenda, most Western European governments were hesitant to generally limit the transfer of dual-use technology because of its consequences for their national export industries. The difference was epitomized by the approval of the British and French administrations to export guidelines for weapons of mass destruction and to an international arms register in the UN Security Council on one hand, while the two governments advocated the reductions of controls on dual-use equipment on the other.

Among the dual-use regulations which were questioned was the COCOM regime. The COCOM had regulated the export of dual-use goods from the Western allies during the Cold War. With the dissolution of the Warsaw Pact, the rationale for COCOM had ceased to exist. Keen to export to the newly opened markets of Eastern Europe, technological companies in the US, Britain and France had successfully pressed their governments to abolish the restrictions. As a result, a temporary revised COCOM list, the New Industrial List, had been agreed in summer 1991. Moreover, a US proposal to enhance cooperation with the former Warsaw Treaty members by replacing COCOM with a less restrictive regime was widely welcomed in Europe where the US had been criticized as too slow on export-control liberalization after the end of the Cold War. In May 1992, an informal COCOM Cooperation Forum was set up to re-negotiate the regime.

Crucially for the German administration's intentions to increase multilateral dual-use controls, COCOM members acknowledged that the need for modern technology in the former Warsaw Treaty countries could not lead to the dismissal of the new danger of proliferation. Moreover, since the revision of the COCOM regime proceeded simultaneously and concurrently with the European negotiations, the German government could seek to use the regime to impose a wider multilateral framework on the EC controls. In coalition with the US government which appeared to support the preservation of at least some of the COCOM controls, the balance of pressures among the COCOM members was marginally more favourable for tighter controls than among Germany's European partners. In order to push the multilateral regime, the German government set up an informal working-group on dual-use goods during its preparation for the G-7 summit in Munich 1992. The group was to continue its work after the world trade summit. However, in spite of the support of the US the German administration was not able to shift the preferences among the COCOM members in favour of a strict follow-up agreement. The proposal for a new successor institution published on 16 November 1992 not only transferred the control of the regime from the international to the national level, but also abolished controls for a large number of goods.

No EC Agreement Before Internal Market

In the European negotiations the German administration failed similarly. In spite of the impending implementation of the internal market, no agreement was reached before the end of 1992. In fact, the British and French representatives vetoed the new proposal which had been suggested in the autumn. To prevent the collapse of dual-use goods controls, the EC Foreign Affairs Council agreed on 21 December to establish interim controls starting 1 January 1993. During the first phase of the decision-making process the German government thus had not only failed to mobilize support for the adjustment of the European and international controls to the German standard, it was itself increasingly under pressure to modify the Foreign Trade Act. When in the course of 1993 the staff from the Economics Ministry relented to the pressure, it strengthened the transnational coalition in favour of a reduction of dual-use export regulations among the European industry and the officials in the economics and trade departments which were engaged in the negotiations. Moreover, additional pressure began to emerge from political actors in Germany.

The implementation of the internal market on 1 January 1993 was used by the transnational coalition among industry representatives and civil servants to reiterate their arguments in favour of less restrictive dual-use export controls. Thus, spokespersons of the European industry association, the Union of Industrial and Employer's Confederation of Europe (UNICE), expressed their concern about licence shopping in the absence of a European dual-use export law. According to the interim agreement, member states were able to maintain their national border controls until a compromise on common regulations for dual-use export controls was reached. Thus, the agreement continued the competitive disadvantages in the export laws across Europe. In fact, with the practical elimination of the COCOM regime, the differences among the dual-use export controls in Europe had become even greater. Since member states seemed unwilling to submit to common EC controls, UNICE representatives suggested a series of bilateral agreements in their place.

In Germany, industry leaders could point directly to the increasingly visible effects of the new German export legislation on the technological industrial base. One example was the decision of the Iranian Defence Industries Organization to move its bureau from Düsseldorf to London. The office had organized the sale of dual-use equipment from over 250 German companies to Iran, all of which had previously been licensed by the government. According to industry representatives 200,000 jobs were in danger due to decreased exports. Although the decline in exports was primarily attributable to international economic developments, industry spokespersons used the argument to demand a lifting of German dual-use export restrictions. In addition, exporters complained that they were discredited by their European competitors as not reliable because the German licensing system could prohibit agreed sales. Since the German controls were not acceptable to other European partners, representatives of the Federation of German Wholesale and Foreign Trade argued that the law should be adjusted to the lower European standards.

Coalition Parties Support Fewer Controls

In February 1993, the industry gained support from a group of 125 CDU/CSU and FDP parliamentarians. In a proposal submitted to the President of the Bundestag, the members of the group urged their own government to revise Germany's export restrictions. The preference change among these politicians came somewhat as a surprise as both CDU/CSU and FDP members in the German Bundestag had supported the strengthening of the controls a year earlier. Moreover, the pressure for a lowering of the dual-use export controls on German parliamentarians amounted to only 5 per cent of their linkages within the network, coming exclusively from representatives of the German industry.[6] However, while the motion was indicative of increasing scepticism over the administration's restrictive export policy among coalition members, the group represented only about a third of the government party members in the Bundestag at the time. It was not until October 1993, that the members of the CDU/CSU and the FDP factions collectively criticized the policy of their ministers on the dual-use issue. It followed the persistent lobbying from industry representatives and the group of the 125 over the summer. Although the parliamentary parties had approved the establishment of the restrictive German dual-use control regime little more than a year ago, CDU/CSU and FDP parliamentarians now unanimously called for the liberalization of German dual-use transfers.

The consequences of the preference change in the CDU/CSU and FDP for the decision-making process were considerable. Specifically, it raised the stakes for the coalition government which relied directly on the support of the CDU/CSU and FDP in the Bundestag for the approval of new export control regulations. In addition, the members of the government parties had close links with the bureaucracy where even lower ranking positions were traditionally held by party members (Mayntz and Derlien, 1989). Officials in the Economics Ministry and the Defence Ministry were particularly sensitive to further pressure since they were already subject to the demand for a reduction of the German dual-use controls of their European counterparts. By utilizing their linkages with civil servants in the two departments, the members of the CDU/CSU and FDP increased the pressure on the staff in the Defence Ministry from 40 to 48 per cent[7] and in the Economics Ministry from 39 to 45 per cent.[8] The probability that the officials from the two departments would acquiesce to the demands of the politicians was thus growing. Indeed, within a week after the coalition parties had collectively expressed their support for a revision of the German dual-use export controls, Reinhard Goehner (CDU), Parliamentary State Secretary in the Economics Ministry, announced that officials in the Economics Ministry supported a revision of the Foreign Trade Act. In line with the German industry associations and the coalition parties, civil servants from the ministry advised that control regulations should be reduced to the European standard.

Foreign Office Resists Mounting Pressure

The sequence of preference changes continued as the staff in the Economics Ministry raised in turn the pressure on the new Economics Minister Rexrodt (FDP) and officials from other departments who were affected by the issue. Among the members of the Kohl Cabinet, Economics Minister Rexrodt was now the most exposed to the pressure for a change in government policy. In total 40 per cent of the actors to whom Rexrodt was linked in the network advocated a revision of the German Foreign Trade Act.[9] However, the pressure was even higher on officials from the Defence Ministry and the Foreign Office. The change of view by their colleagues in the Economics Ministry meant that now respectively 52 and 42 per cent of the Defence Ministry and Foreign Office contacts in the network favoured a reduction of dual-use export controls.[10] Moreover, due to their boundary position between national and international actors, officials from the Economics Ministry were able to provide a transnational bridge between the intergovernmental coalition for a revision of the German dual-use controls among their European colleagues and domestic actors in the European foreign policy network. Over the following months this position allowed Economics Ministry staff to link the pressure from both international and national actors on the German administration.

Only shortly after these changes, an announcement of another Parliamentary Secretary in the Economics Ministry, Heinrich Kolb (FDP) raised expectations about an impending change in government policy. Kolb stated that the government had recognized the need to examine the consequences of its national export legislation. Moreover, according to Economic Ministry officials, the administration had recognized that it was preferable to accept lower standards than to prevent the harmonization of the dual-use controls in the EC. However, cabinet ministers remained intent on bargaining for the highest standards possible. The strict line over dual-use export controls was in particular based on resistance from Foreign Minister Kinkel and officials in the Foreign Office who had born the brunt of the criticism over the German arms export scandals during the early 1990s and were, therefore, critical of less restrictive controls.

Foreign Office staff continued to oppose the weakening of German dual-use export controls during 1993, although they were at the centre of considerable pressure from their European colleagues, from the Economics Ministry and from the members of the CDU/CSU and FDP parliamentary party who together accounted for 42 per cent of their relations in the network. They formed a close alliance with Foreign Minister Klaus Kinkel who was especially reluctant to modify his position on the issue. The ability of Kinkel to maintain his opposition to a reduction in German dual-use regulations was enhanced by the fact that he was exposed to one of the lowest degrees of pressure in the German administration at 34 per cent.[11] However, the balance in favour of retaining the existing German dual-use export laws was shifting among other government ministers. In August 1993, Defence Minister Rühe tentatively supported calls from the German industry to reestablish its ability to compete on the international technological market. Since Rühe had no direct authority over the issue, however, he continued to adhere to the government position that German dual-use export controls would not be reduced.

Limited Progress in EC Negotiations

In the meantime, the European negotiations made first progress on the basis of a Belgian proposal which suggested a distinction between the general framework of the controls set by a Commission regulation and the content of the lists which would be flexible and under constant review under the Common Foreign and Security Policy decision-making process (Cornish, 1997: 40). On the basis of this compromise, the European Commission submitted a new draft in autumn 1993. The question of a country list remained open. The request of the German representatives for an obligatory catch-all clause, similar to §5c of the German Foreign Trade Act, was again rejected. However, on condition that it was restricted to weapons of mass destruction and carrier missiles, the governments of some member states appeared to consider supporting the clause. In particular, within the Italian administration there was initial encouragement for tighter regulations. Thus, German negotiators continued to demand a catch-all-clause. Nevertheless, the pressure from the German administration had weakened since officials from the German Economics Ministry, who led the German representation, had publicly expressed their support for less restrictive common controls. The German delegation appeared content to seek to safeguard existing national dual-use export regulations. To this purpose they used the German veto position in the intergovernmental negotiations. Specifically, German Economics Ministry representatives secured an opt-out formula which allowed member states to maintain or implement stricter regulations. Amongst others, the opt-out formula would apply to the control of technological knowledge and services which was regulated in the German Foreign Trade Act.

The revision of the COCOM regime which was concluded at the same time equally failed to establish more comprehensive multilateral dual-use transfer regulations. Meeting in The Hague on 16 November 1993, COCOM member states agreed to dismantle the old regime and to replace it with a new institution by the end of 1994. During the interim period only a core list, the so-called Interim List, was to be controlled which gave more discretion to the member governments. The weakening of the COCOM regime was a direct result of a change of policy in the US. Initially, US Secretary of State Warren Christopher had argued that non-proliferation was the most important challenge to the US and Europe and had praised the efforts of the EC to regulate its dual-use exports. However, the new US administration was more concerned about its export figures than about proliferation. President-elect Bill Clinton announced that it was his intention to encourage the research and development of dual-use goods through a range of government incentives.

First Preference Changes among German Ministers

In the meantime, the domestic support for a revision of the German export laws continued to mount. CDU/CSU and FDP parliamentarians strategically lobbied Minister of Economics Günter Rexrodt and Foreign Minister Klaus Kinkel. Both actors were not only under pressure from the transnational coalition that had emerged among officials from the German and European Economics Ministries, but also had the direct authority over the issue in the Cabinet. In a direct appeal to the two ministers, CDU/CSU foreign trade spokesman, Peter

Kittelmann, asked the government to reconsider its position on dual-use export regulations. With reference to the opposition from within the Foreign Office to a revision of the Foreign Trade Act, Kittelmann demanded that the Economics Ministry should recover its 'leadership' on the issue. According to CDU/CSU parliamentarians the progress towards common European dual-use export controls should not be prevented by the Foreign Office.

Given the combined international and domestic pressure on Rexrodt, the strategy of the CDU/CSU and FDP parliamentary parties soon paid off. By mid-November, the Economics Minister publicly expressed his support for a review of the Foreign Trade Act. Rexrodt was, thus, the first cabinet minister to abandon the existing German dual-use transfer policy. Rexrodt's support was a crucial success for the growing coalition in favour of a revision of the German dual-use export regulations. With the support of the minister, the coalition had not only gained direct influence over other cabinet ministers, but also a central voice in the Cabinet itself. Due to the preference change of the Economics Minister the pressure on all other members in the Cabinet increased notably. Defence Minister Rühe was now subject to the highest pressure with 38 per cent of the actors to whom he was linked in the network supporting the reduction of German export controls.[12] As has been noted above, Rühe had expressed his doubts over the tight German regulations already in August, but had been forced to adhere to the official government line on the issue. Following Rexrodt's preference change, Rühe soon came out in support of Rexrodt's position. In the Cabinet, Foreign Minister Kinkel again prevented a change of policy. The Foreign Minister voiced his concern that a reduction of dual-use controls would encourage the arms build-up in volatile regions such as the Middle East. However, the preference changes of his two Cabinet colleagues put Kinkel increasingly on the defensive since now 42 per cent of his contacts in the network called for a revision of the Foreign Trade Act.[13]

The series of preference changes in autumn 1993, thus, ended with a stalemate in the German Cabinet. Although the combined pressure from the international community and an increasing number of German actors in the European foreign policy network had brought the government to the brink of a revision of the Foreign Trade Act, it was only when the pressure was maintained during the following year that the German administration began to make gradual changes in its dual-use export policy.

Government Parties and Industry Use Indirect Pressure

Due to their failure to achieve a change in policy, CDU/CSU and FDP parliamentarians, as well as industry leaders, expanded the use of their relations within the European foreign policy network during the winter of 1993-94.

While the parliamentarians initially concentrated on their direct relations with cabinet ministers and civil servants, the members of the CDU/CSU and FDP now increasingly also employed their relations with the media, through the Bundestag and its committees to exert indirect pressure on the Cabinet. As part of this strategy which sought to broaden the support for a reduction of dual-use export controls within the network rather than focus directly on decision-makers, members of the CDU/CSU parliamentary party submitted several memoranda drafted by its foreign policy working-group to cabinet ministers and officials. In the documents, the party's foreign policy and economic experts appealed to the government to proceed with the harmonization of European export regulations even if this entailed the reduction of German dual-use goods controls. Specifically, they urged ministers to accept concessions on the equipment and country lists. Another part of this strategy involved the distribution of these proposals to the media in order to increase public pressure.

In addition, members of the coalition parties began to use their parliamentary majority in the Bundestag to collectively exert pressure on the administration. Since the coalition members hesitated to challenge the government directly in the Bundestag as a matter of party solidarity, the CDU/CSU parliamentary party decided to convene a Bundestag hearing of industry representatives at the beginning of December. The hearing provided spokespersons of industry associations and major companies with an opportunity to exert direct pressure on the relevant politicians. Moreover, parliamentarians were able to use the evidence from the hearing to support their demands for less restrictive dual-use export controls. During the hearing, representatives from the main German industry associations reiterated their concern that the current regulations undermined the ability of the German industry to compete in the world market. In particular, the representatives attacked the catch-all clause and the control of knowledge and service exports which remained under national control due to the EC's opt-out agreement.

At the same time the industry employed its links with the media to exert indirectly pressure on the government for a revision of the Foreign Trade Act. Speaking to Germany's main national newspaper, the conservative *Frankfurter Allgemeine Zeitung*, representatives of the Federation of German Chambers of Commerce and Industry (DIHT) and the German Association of Machinery and Plant Manufacturers (VDMA) repeated their demands that German laws were adjusted to the average European standard. The industry associations' spokespersons claimed that the existing controls contributed to or even were the cause of the increasing difficulties of the German armaments and manufacturing industry and the rising unemployment in the sector.

The statement indicated a shift in the debate which was marked by the increasing demand to relax not only controls for dual-use equipment, but also for armaments exports. The problems of the arms industry were symbolized by DASA in Lower Saxony, one of the German Länder that had been suffering most from the conversion of the arms industry since the late 1980s. Although German labour unions, as represented by the German Labour Union Association (DGB), generally tended to be in favour of strict export controls, union representatives of DASA met personally with ministers and parliamentarians to ask for a relaxation of German export rules on dual-use goods in order to rescue the company's future. The governments of Länder with a high proportion of armaments and manufacturing industry were especially susceptible to the warning that unemployment in the sector was rising. Moreover, the Bavarian Minister for Economics and Transport, Dr Otto Wiesheu (CSU), argued that Germany also had to harmonize its weapons export controls to recover its ability to co-operate in multinational armaments projects and to maintain Germany's influence in NATO. Even SPD-governed states such as Lower Saxony and Schleswig-Holstein, which had a high percentage of armaments and shipbuilding industry, were divided over the tight German export controls. However, the representatives of the Länder in the Bundesrat had no direct authority over national export legislation.

Control Debate Extended to Weapons

Although the evidence available did not support the argument that the German technological base and employment in Northern Germany could only be maintained authoritative influence of the Foreign Office. The extension of the dual-use control debate to armaments exports, presented another attempt by industry representatives to change the conditions of the policy process by strategically shifting the decision-making authority over the issue to actors who were more favourable towards a revision of the German Foreign Trade Act.

While the issue of dual-use transfers had been exclusively under the authority of the Economics and Foreign Ministries, the question of armaments exports led to a greater involvement of the Ministry of Defence in the decision-making process. Crucially for the aims of the industry, Defence Minister Volker Rühe supported its demands on both the dual-use and armaments export control issue. In a letter to Chancellor Kohl, Defence Minister Rühe contended that it was in Germany's security interest to maintain a defence industrial base and to be able to collaborate in international armaments projects which were currently prevented by differences between the national export regulations in Europe. Like industry representatives, the Defence Minister criticized Foreign Minister Kinkel and the Foreign Office as the main obstacle to the required revision of Germany's export rules. Only recently Foreign Minister Kinkel had blocked the licensing of arms transfers to Taiwan in the Cabinet's Federal Security Council.

Shortly after the Bundestag hearing, CDU/CSU and FDP parliamentarians sought to extract further concessions from cabinet ministers at the institutionalized Coalition Meeting. At the Coalition Meeting, the leaders of the CDU/CSU and FDP factions as well as the leading cabinet ministers were represented and able to exert direct pressure on each other. Given the unanimous support for a revision of the Foreign Trade Act among the CDU/CSU and the FDP parliamentary parties, the pressure on the Coalition Meeting in favour of a revision of the German dual-use controls was at 47 per cent[14] significantly higher than in the Cabinet at 35 per cent.[15] In fact, the Coalition Meeting was subject to the third highest degree of pressure in the foreign policy network following the Bundestag and officials in the Defence Ministry. As a result of the pressure from within the coalition, Chancellor Kohl and the leading members of the CDU/CSU and FDP conceded that the government would show a greater willingness to reconsider the German dual-use legislation in the European negotiations in order to allow for a settlement. However, Chancellor Kohl continued to urge for a European harmonization on the basis of the German legislation - as far as possible.

To discuss the terms of new German export rules, the government set up a working group of officials from the Economics Ministry, the Defence Ministry and the Foreign Office. In response to the lobbying from the industry and the Defence Ministry, the ministries were, for the first time, ordered also to consider the effects of the dual-use sale regulations on employment and the maintenance of the defence industrial base. The talks concentrated on attempts to speed up the licensing of collaboration between German and European technological companies. With respect to international collaboration in technological developments, the issue of linking dual-use export controls with weapons transfer regulations reemerged. In particular, an alliance between industry representatives and officials in the Defence Ministry favoured a regulation that combined the Foreign Trade Act with the War Weapons Control Act in a comprehensive and less restrictive form. However, the attempt to link the two issues failed. While the Cabinet was prepared to compromise on the controls of dual-capable technology, the majority of cabinet ministers opposed the liberalization of armaments sales. Foreign Minister Kinkel and Minister of Economics Rexrodt proclaimed publicly that the War Weapons Control Act of 1982 would not be changed.

Armaments Collaboration Eased

The ability of the Cabinet to resist demands for the coupling of dual-use with arms exports was enhanced by the fact that CDU/CSU and FDP coalition parties were internally split on the question of liberalizing weapons transfers. Foreign policy spokesman of the CDU/CSU, Karl Lamers, published a memorandum in which he advocated the reduction of German armaments export regulations along with common European regulations for dual-use equipment. Conversely, CDU/CSU deputy faction leader Johannes Gerster contended that illegal armaments exports should be fought first, before more generous export licensing was considered. Several leading members of the CDU/CSU faction stated their objections to Lamer's memorandum. Germany's labour unions were similarly divided on this question. On one hand, Klaus Zwickel, the leader of the national federation of the metal workers, IG Metall, took the view that armaments exports were dangerous and without future. On the other, representatives of the Rheinmetall union contended that it was necessary to maintain a national defence industry as a contribution to the European security system.

The Cabinet reacted to the competing pressures by distinguishing between licensing procedures for collaborative armaments projects among the members of the EC and NATO and armaments exports to third countries. While cabinet ministers did not want to appear to promote global proliferation, the former were supported by Defence Minister Rühe and civil servants in the Defence Ministry and the Economics Ministry. Initially, Defence Ministry officials had demanded that the German defence industry should not only be able to collaborate in European armaments projects for the home market, but also for international exports. However, ministers agreed to maintain Germany's tight weapons export regulations. On 13 January 1994, the German Cabinet resolved the issue with a law designed to enable the cooperation and the sharing of information under a European armaments export regulation.

Legal Amendments Indicate German Policy Shift

In the European negotiations over the dual-use export controls, the German representation had resigned to the fact that it would not be able to achieve stricter regulations than had been agreed in the compromise of October 1993. Although the European Council of Ministers was expected to approve the draft for common European dual-use export regulations in May or June for implementation on 1 January 1995, an agreement with the governments of France, Britain and Italy over which countries should be banned from dual-use exports was still missing. Nevertheless, the German Cabinet approved a series of amendments to the Foreign Trade Act which were designed to pave the way for the common European dual-use controls. The first amendment of the Foreign Trade Act on 28 February 1994, primarily made terminological corrections in response to the internal market. It also allocated the authority over the implementation of the common EC regulations to the appropriate national departments.

Members of the Social Democratic (SPD) opposition party criticized the amendments and demanded that the loopholes which were created by the common European regulations be closed by additional national controls. Moreover, SPD parliamentarians argued that the government was giving the wrong signals to Brussels. Members of the Greens and the Party of Democratic Socialism (PDS) supported them. However, in spite of their unified opposition to the relaxation of German export controls, the opposition parties were not able to prevent the impending policy change. Following the extension of the transnational coalition in favour of a reduction of the German dual-use controls from civil servants to two cabinet ministers as well as the CDU/CSU and FDP parliamentary majority, the pressure on the German Cabinet was considerable, amounting to 35 per cent of its linkages within the network.

While the previous analysis has suggested that the Kohl Cabinet had been able to withstand the collective pressure from nearly all European governments due to the veto position of the government in the international negotiations, the combined demands from national and international actors began to compromise the Cabinet's resistance. Due to the support for a revision of the German dual-use goods controls among domestic actors, cabinet ministers began to consider the lifting of its national dual-use controls even beyond the requirements of the European compromise. In talks with the Saudi-Arabian leadership in early February, Minister of Economics Rexrodt suggested that a revision of the government's policy on dual-use exports was imminent. Rexrodt envisaged specifically a reduction of Germany's 'H' country list. To sustain the pressure on the Cabinet, the Economics Minister and officials from the Defence Ministry sought to use transnational and international linkages with their European colleagues. In particular, Rexrodt attempted to transfer the issue of dual-use exports to international organizations where he could draw on their support for less restrictive regulations by redefining the authority over the issue. Thus, Minister of Economics Rexrodt argued that the Germany Foreign Trade Act needed to be reduced to comply with the higher authority of international standards set by the multilateral dual-use control regimes of which Germany was a member.

COCOM Replacement Reduces Controls

Incidentally, multilateral dual-use export controls had been significantly cut back with the abolition of the COCOM regime in the previous year. Moreover, although an *ad-hoc* working group had been set up at a meeting in Wassenaar, Netherlands, in November 1993 to draft a proposal for a COCOM successor regime, it soon became clear that the new regime would be much weaker than its predecessor. The first round of negotiations failed to reach an agreement and a second round had just opened on the bureaucratic level in March 1994. The French government, in particular, opposed tight controls. With reference to its national sovereignty, the French delegation vetoed the full control of conventional armaments technology. The US administration, which had initially resisted the weakening of the COCOM controls, had also relinquished its stance on dual-use exports. Thus, the Clinton administration issued a general licence dual-use exports to the former Soviet Union and China in order to ease exports. The move prompted fears among the European armaments industry that the US would take advantage of the weak multilateral regimes, while European exports were curtailed by the emerging common export regulations.

The proposal for a COCOM successor regime, which was unveiled in April, confirmed fears among proliferation experts that national differences of interest would prevail. Although the member states had agreed on a set of common rules, the draft envisaged that licensing would be conducted nationally. Moreover, country controls were reduced to very few states such as Iran, Iraq, Libya and North Korea. While a decision on the draft was still outstanding and, in fact, not reached until July 1996, Economics Minister Günter Rexrodt immediately suggested the reduction of Germany's H-list from 33 to eight or ten countries to conform with the new arrangement. However when questioned by the press about the position of the German government on his proposal, Rexrodt had to admit that his initiative had not been co-ordinated with his Cabinet colleagues.

Officials from the Defence Ministry pursued the same strategy of transferring the issue of export regulations to other authorities. Following the officials' failed attempts of December 1993 to reduce German armaments export controls in line with the revision of its dual-use transfer regulations, Defence Ministry officials now sought to utilize the opposition to strict multilateral controls among its European partners for their cause. In a working paper published on 24 May, the civil servants proposed the creation of a European Armaments Agency with full authority over European arms exports. The paper had been designed as the basis for negotiations about a West European Armament Group in the WEU. Although the officials from the Defence Ministry insisted that the organization was not intended to undermine the German armaments export legislation, it would obviously limit the scope for national exceptions. Crucially, the paper suggested changing article 223 of the Treaty of Rome in the planned revision of the Maastricht Treaty in 1996 to transfer the decision-making authority on armaments questions to the European level. The plan was criticized by the SPD opposition as 'dangerous and stupid'. SPD members pointed out that a European Armaments Agency could be used to end national export controls. Thus, the agency would threaten the clause that had been secured in the negotiations over common European dual-use export regulations which allowed for additional national controls.

However, in May 1994, the combined pressure from officials in the German Economics and Defence Ministry and their European colleagues in NATO, COCOM and other multilateral export control regimes showed some success. As consequence of the negotiations about the COCOM successor regime, officials from the German Economics Ministry announced that the government was relaxing the rules for armaments cooperation with members of the Organization for Economic Co-operation and Development (OECD). The new regulation allowed German companies to export sensitive technology to OECD members without a licence if the goods contributed to not more than 20 per cent of the finished product. Weapons of mass destruction and missile technology, however, were exempted from the licence. The Cabinet approved the decision on 8 June 1994.

Transnational Coalition of European Industry

Encouraged by the success of the transgovernmental coalition among the civil servants from the European Economics and Defence Ministries, representatives from the German technological industry also mobilized their transnational relations in order to exert pressure on the German administration. As the European negotiations were drawing to a close, the industry used its international European associations to lobby against the stipulations which allowed EC members to maintain additional national export controls. 'It would be difficult, if not impossible, for a major company to contemplate collaboration with another company unless that company exists in a country where the government has accepted common principles of exporting policy', concluded a spokesperson of the European Defence Industry Group (EDIG).[16] Representatives of UNICE, the European association of businesses, demanded a quick resolution of the negotiations in Brussels in order to create an equal competitive environment.

However, industry representatives were less successful than the civil servants. At their meeting of the European Council on Corfu, 24-25 June 1994, European foreign ministers approved of the draft for common dual-use export controls as it had been agreed over the past year (Cornish, 1997: 40). The general framework of the regulation was to be implemented in the form of a directive by the Commission. But the specific content of the contentious equipment and country lists would be open to constant review by the member states under the joint action decision-making institutions of the European Common Foreign and Security Policy (CFSP). The final package contained lists of chemical, biological, nuclear and missile products which would require an export licence for transfers outside the EC. Their content widely matched the German lists. However, unlike the German Foreign Trade Act, the document did not include the transfer of sensitive knowledge and services. A regulation similar to the German catch-all clause §5c which controlled equipment that to the knowledge of the exporter was intended for armaments production had also not been acceptable to the other EC members. To compensate for these omissions, an opt-out formula allowed member states to maintain stricter national regulations where considered essential for national security. Although this formula should have paved the way for an agreement on the country lists by leaving the decision over the scope of the controls with the member states, the member governments continued to deliberate the content of a common European country list.

EC 'Deadlock' Justifies Country Cuts

Whether officials from the German Economics Ministry actively prolonged the discussions about the country controls does not emerge from the data. However, the continuing debate had a crucial impact on the decision-making process as it reasserted transnational and international pressure for a reduction of the German H-list during the summer and autumn of 1994. Minister of Economics Rexrodt used the deadlock in the European negotiations on the country list question to insist in the German Cabinet on a reduction of the German country controls. Specifically, Rexrodt wanted to restrict Germany's national exemptions from 32 to six states, including Iran, Iraq, Libya, North Korea and Syria. Rexrodt's fellow FDP party member and Cabinet colleague Foreign Minister Klaus Kinkel and officials from the Foreign Office opposed the plan. Kinkel justified his objections with the argument that a reduction of the H-list would permit exports to Algeria, Angola, Pakistan, Cuba and Vietnam which were known for their human rights violations. In an official statement, Foreign Office civil servants rebutted the suggestion with the comment that with regard to the country list an agreement had not been reached. However, Kinkel and his staff were under high pressure due to their boundary position which exposed them to pressure from both national and international actors. Including the united coalition among the other European

Foreign Ministries and his FDP party colleagues, 42 per cent of Kinkel's contacts in the network demanded a reduction of the country controls.[17]

Support for the Foreign Minister and the Foreign Office staff on the question of the country controls came only from members of the SPD, the Green and the PDS opposition parties. However, the influence of the opposition members on the government was limited. Conversely, opposition parliamentarians were dependent on cabinet ministers and civil servants for expertise and regular information about the international negotiations. At best opposition members could exert indirect pressure via the media and the electorate during the ongoing general election campaign. An opportunity to direct public attention to the impending change of government policy arose during a meeting between EC foreign ministers and governmental representatives of ASEAN states. Using the meeting as a platform, SPD members urged the Cabinet in the media to refrain from allowing arms transfers to the region. In their election manifesto the SPD promised to restrict armaments exports to members of NATO and the EC, if elected to government.

However, during the election campaign the question was not promoted as a central issue because of its association with job losses in the technological industries. Since even the German labour unions were divided over the question of armaments exports, the SPD was best advised not to raise the issue with its grass roots supporters. In fact, only 39 per cent of the electorate in West Germany favoured the banning of armaments transfers. In the former East, where the PDS had its main base, 52 per cent of the electorate supported the abolition of weapons exports. The PDS wanted to prohibit all arms exports, as did the Greens. However, when the Kohl coalition government won a third term in office in the general election of 16 October 1994, the only chance of the opposition members to prevent the revision of the German dual-use export controls by gaining control over the decision-making process was lost.

German Dual-Use Control Law Changed

In early December, a Coalition Meeting between the Chancellor, key cabinet ministers and the leaders of the CDU/CSU and FDP parliamentary parties decided to reduce German dual-use export controls. The decision was no surprise. The Coalition Meeting had been under considerable pressure during the autumn from 53 per cent of the actors to whom it was linked in the network. In particular, the members of the CDU/CSU and FDP parliamentary parties, Economics Minister Rexrodt and Defence Minister Rühe favoured a revision of the Foreign Trade Act.[18] Although not a formal institution of the German policy process, the decision of the Coalition Meeting had important consequences. Practically, the Coalition Meeting served to negotiate all controversial issues in the coalition government. As such, it had direct influence over the Cabinet. With the approval of the Meeting to a reduction in the German dual-use export controls, the pressure on the collective Cabinet was raised from 40 to 45 per cent.[19] Moreover, the Coalition Meeting increased the pressure on both Chancellor Kohl and Foreign Minister Kinkel also to 45 per cent, that is nearly half of the actors to which both were linked within the European foreign policy network supported the revision of the German dual-use export licensing conditions.[20] On 9 December 1994, the Cabinet approved the amendment of the German Foreign Trade Act.

Following the Cabinet decision, officials from the Economics Ministry and the Foreign Office reduced the 'H' country list from 32 to 9 states in long and protracted negotiations between the two ministries. Removed from the list were among others Egypt, China, Pakistan, India, Angola, Algeria, Vietnam, Yemen, Cambodia, Lebanon and Taiwan. Foreign Office staff had originally wanted to include ten new countries which were affected by civil war, such as Georgia and Tajikistan, or which were subject to international embargos, including Armenia, Azerbaijan, Haiti, Nigeria, Ruanda, Sudan and Zaire. However, after the decision of the Cabinet, the ability of the Foreign Office to influence the content of the country list was very limited.

When it came to justifying the reduction of the country list in public, however, the German government was at a loss. Cabinet ministers could hardly admit that the revision of government policy had been a consequence of persistent and strategic pressure from the German armaments and export industry. Instead, Economics Minister Rexrodt falsely claimed that the amendment had become necessary due to the European agreement on common dual-use export regulations. As the German European Parliament Member Jannis Sakkellariou (SPD) correctly pointed out, however, the shortening of the country list was by no means required by the EC regulations. The British government, for instance, maintained a list of 44-45 countries for special controls.

EC Agreement Reached

On 19 December 1994, shortly after the policy reversal of the German government, the European Council approved the common regulation for the control of exports of dual-use goods. Differences between the German controls and the common European regulation pertained to the lists of dual-use equipment which were less comprehensive than the Foreign Trade Act in the area of chemical and biological goods. However, with the support of a number of EC member states, the German government had achieved the inclusion of a catch-all clause for nuclear, biological and chemical weapons. Due to the opposition of the French administration, the catch-all clause exempted conventional technology. Moreover, exporters only required a licence if they had been informed by the authorities of the military use of their equipment or if they had 'positive knowledge' that it was intended for military purposes. Yet, the European regulation included several stipulations which enabled member states to maintain stricter controls if they chose to do so. In line with these, the German government retained its controls for dual-use goods which could be converted to conventional weapons. Furthermore, Germany required licensing for dual-use equipment which was to be exported to countries on its new country list, the X-list. National regulations for the transfer of sensitive know-how and technical services also remained in force.

Assessment of the Hypotheses

In order to assess the hypotheses proposed by multilevel network theory, this section draws together the evidence presented in the previous analysis. Specifically, it uses four indicators to examine the degree to which the hypotheses were corroborated in the making of European dual-use controls. The first indicator concerns the likelihood of a preference change at specific degrees of pressure. The second regards the differences in the distribution of four behavioural categories, i.e. no change of preferences, change of preferences, unclear or undecided preferences and the blocking of a preference change in collective decision-making units, in relation to the pressure to which the actors were exposed. The third indicator is the average degree of pressure for each of the four behavioural categories. The fourth measure analyses when actors changed their preferences during the decision-making process and whether the timing of a preference change was related to an increase of pressure on the respective actor. In addition, to providing an assessment of the hypotheses, this section analyses whether the empirical findings can suggest new or more specific hypotheses concerning the relationship between changes in actors' foreign policy preferences and the degrees of pressure to which they were subjected.

The first proposed measure applies to the first hypothesis of multilevel network theory which suggests that rising degrees of pressure can be associated with a higher frequency of preference changes among the affected actors. Crucially for a positive assessment of the theory, it is broadly confirmed by the findings in this case as displayed in Figure 3.1. The figure shows the number of actors who changed their preferences out of all actors who were exposed to pressure within 5-per cent intervals between zero and a hundred per cent.[21] Although the relatively small number of instances in a single case study does not lead to a continuous curve, the findings demonstrate that higher degrees of pressure were nearly always associated with an increase in the proportion of preference changes.

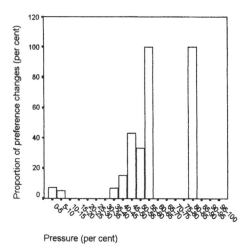

Pressure (per cent)

Figure 3.1 Proportion of preference changes

Further support for the first hypothesis can be derived from the second indicator, namely the distribution of the four behavioural categories - no change of preference, unclear or undecided, change of preference and blocked preference change - across the full range of pressure between zero and one hundred per cent. The first category which applies to those actors who did not change their policy preferences is shown in Figure 3.2.

Initially it appears surprising that the distribution shown in the figure peaks twice, first at 5-10 per cent and again at 30-35 per cent. The first hypothesis would lead to the expectation that the number of actors who maintain their preferences during a certain point of the decision-making process decreases steadily the higher the degree of pressure to which they are exposed. Visually, this would be represented by a diagonal distribution from many instances of 'no change of preferences' at low degrees of pressure to few instances at higher degrees of pressure.

However, the divergence can easily be explained by the choice of the research period. Since the case study examined the decision-making process during the months in which the German government came under increasing pressure to reduce its dual-use controls, very few actors were under no or little pressure. A different selection of the research period, beginning at a time when none of the actors seriously pressed for less restrictive dual-use export regulations, for instance in 1989, would probably increase their number.

Nevertheless, the first hypothesis is in so far validated as the number of instances in which actors maintained their original preferences falls progressively after the second peak. In fact, the frequency of the category 'no change of preferences' falls to zero at a pressure of 55 per cent, suggesting that there might be a threshold beyond which actors were not able to resist further pressure for a modification in their policy preferences.

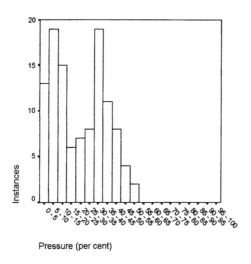

Pressure (per cent)

Figure 3.2   No change of preferences

The second behavioural category which concerns actors who displayed unclear preferences or who were divided over which policy to pursue in Figure 3.3 provides further evidence for the first hypothesis since it peaks at somewhat higher degrees of pressure than the number of actors who did not change their preferences. Typically undecided actors were subject to pressures between 30 and 45 per cent. The findings thereby seem to indicate an intermediate phase in which actors are considering whether to change their policy preferences in response to the pressure from other actors in the network.

In addition, the six instances of unclear preferences at pressure of 65-75 per cent appear to indicate that being unclear about their preferences enables actors to resist higher degrees of pressure. However, since each of these instances applied in this case to the same actor, namely the German Bundestag, the findings cannot be generalized.

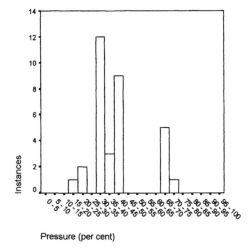

Pressure (per cent)

Figure 3.3 Unclear/undecided preferences

Moreover, the fact that the Bundestag's preference was codified as 'unclear or undecided' was a consequence of the formal support for the government among the members of the coalition parliamentary parties. If the parties had decided to approve of an amendment of the Foreign Trade Act before a Cabinet decision, this would have been an open rebellion and could have led to a vote of no confidence. Although the CDU/CSU and FDP parliamentary parties were unanimously in favour of a reduction of dual-use controls, they chose not to challenge their government in the Bundestag. Nevertheless, as party members, they repeatedly and publicly criticized the government's policy.

The number of actors who did change their preferences during the decision-making process in Figure 3.4 also supports the first hypothesis as it reaches its peak at 40-50 per cent, i.e. higher degrees of pressure than both the actors who did not change their preferences and those who were unclear or undecided. Only the preference changes among the members of the CDU/CSU and FDP parliamentary parties at 5 per cent pressure, contradict the first hypothesis. An alternative explanation for this behaviour could involve degrees of power to account for the presumed close relationship between the German industry and the coalition parties. Thus, in particular the FDP perceives itself as the party political representation of the German industry. However, as has been argued in the exposition of multilevel network theory in the previous chapter, such an approach would incur serious theoretical and methodological problems. Since the error margin encountered this way is rather narrow, it appears more fruitful to treat the two instances as exceptions from a generally highly valuable hypothesis.

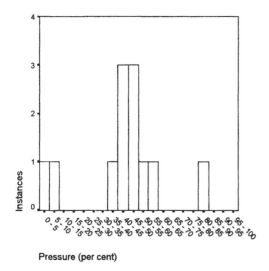

Figure 3.4   Change of preferences

The use of a blocking strategy which pertains to the second hypothesis was employed by none of the actors in this case, primarily because of the nature of the issue. Thus, the maintenance of high export standards in Germany required a positive decision for a policy change which included the establishment of tight international controls and prevented the use of a blocking strategy by the German government. Conversely, since the introduction of the internal market threatened to undermine German dual-use controls by default, the German government had to lobby for support in favour of strict European regulations. A similar situation applied to the negotiations for a COCOM successor regime. The nature of the regime as an intergovernmental and voluntary agreement did not allow for a German veto over the dismantling of the controls. Again, the government was conversely forced to mobilize support for a new multilateral control regime.

Domestically, however, the impending policy reversal could have been blocked by a majority in the Bundestag. Yet since the coalition government's parliamentary parties were among the strongest and earliest advocates of less restrictive dual-use export controls, a veto was not considered by them. The members of the three opposition parties in the Bundestag, which were opposed to an amendment of the Foreign Trade Act, were on the other hand unable to veto the revision of German dual-use export regulations due to their minority position within the Bundestag.

The third indicator which concerns the average degree of pressure within each of the four behavioural categories underscores the findings presented above. Thus, it shows that there is a significant difference between the average degree of pressure at which actors were able to maintain their original preferences at 22 per cent and those at which they modified them at 42 per cent. The average of the pressure at which actors were unclear or divided at 41 per cent is slightly below the average pressure at which actors change preferences, providing further indication for an intermediate phase of reconsideration and insecurity at an intermediate range of pressure.

Table 3.1   Descriptive statistics

Preference changes	Number of instances	Range of pressure	Minimum pressure	Maximum pressure	Average pressure
No change (NC)	112	52%	0%	52%	22%
Unclear or undecided (U)	33	58%	15%	73%	41%
Change (C)	12	77%	5%	82%	43%
Blocked (B)	0	-	-	-	-

Finally, the timing of the preference changes further helps to substantiate the value of the first hypothesis which links policy preference changes to the degree of pressure exerted over network actors. In particular, this measure reveals that nine of twelve actors changed their policy position immediately after the pressure on them had increased. These actors included the members of the European Commission, staff from the German Economics Ministry, the German Economics Minister Rexrodt, Defence Minister Rühe, staff from the Defence Ministry, the members of the Coalition Meeting, Chancellor Kohl, the Cabinet and Foreign Minister Kinkel.

Among the three actors who did not immediately respond to growing pressure for a modification in their preferences, the Foreign Office staff responded in the second phase after an increase. The belated preference change among Foreign Office officials matches their long-standing dedication to the cause of tight dual-use controls. However, the deferment of their preference reversal by merely one phase does not seriously challenge the close correlation between pressure increases and the timing of preference changes. The other two exceptions pertain again to the members of the CDU/CSU and FDP parliamentary parties who modified their policy preferences although they were not subject to increasing, or indeed substantive, pressure from other actors within the multilevel foreign policy network.

Conclusion

The beginning of this chapter raised the question how multilevel network theory can help us understand the interactions within the European foreign policy network and their impact upon the foreign policies of the EC and its member states. In order to answer this question, this chapter has presented a detailed analysis of the making of European and German dual-use export controls between 1992 and 1994. Furthermore, it has examined to what degree the two hypotheses proposed by the theory were corroborated by the empirical findings. Combining both elements, several conclusions can be drawn.

The first observation concerns the structure of the multilevel European foreign policy network and the range of national, transnational and international actors which are involved in European foreign policy decision-making. Specifically, the preceding case study shows how national actors influenced decisions within the EC and vice versa. However, the preceding analysis also reveals that the European foreign policy making process was not restricted to the EC and its member states. A broad variety of international institutions, such as the COCOM regime and the Australia Group, played a role in defining which foreign policy options were available with regard to the multinational control of dual-use transfers. In addition, the United States administration influenced the process both by its support for a common European control regime and its willingness to reduce controls within a COCOM successor institution. Moreover, multilevel network analysis demonstrates that domestic actors were able to exert considerable pressure on national and international decision makers. The case study thereby reveals that even negotiations within the European Community are crucially determined by the national, transnational and international relations of the EC and its member states within the multilevel European foreign policy network.

A second set of conclusions regards process of European foreign policy decision-making. The most central observation is that process matters. The case study and the subsequent quantitative assessment of the hypotheses demonstrate that most actors changed their preferences concerning dual-use export controls as the result of a sequence of preference changes which over time increased the degree of pressure to which they were exposed from other members of the multilevel European foreign policy network. The formation of a winning coalition in favour of less restrictive dual-use export laws in Germany can thus be explained by a form of domino effect. Whenever one actor changed his or her policy preference, he or she in turn raised the pressure on other members of the network.

However, the preceding analysis also illustrates that the structure of the network determines which actors are able to influence others and contribute to the formation of a winning coalition. German ministers and civil servants played a particularly important part in this coalition-building process because these actors held so-called 'boundary roles' between national and international actors. As a consequence, they were able to form a coalition with actors in both arenas in favour of a reduction of German dual-use export controls.

Yet, the case study suggests that there are some misconceptions about the gatekeeper function of boundary roles. In particular, the evidence from this case study suggests that gatekeeper or boundary roles may be able to use their national and international relations to promote their own policy preferences, but they are also more vulnerable because they are linked to more actors who can exert pressure on them, as the preference change of staff from the German Economics Ministry which led the German delegation in the EC negotiations shows. Furthermore, the case study indicates that direct transnational linkages among private actors such as German industry representatives and the Commission or international industry associations can help to circumvent the gatekeeper position of boundary actors.

Finally, multilevel network analysis provides new insights into the question who determines national and international European foreign policies within the context of the European Community. It suggests that neither domestic nor international actors dominate in the European foreign policy decision-making process. Contrary to the argument made by the German government and perhaps first impressions, the decision of the German administration to cut back its national dual-use controls cannot be fully explained by the European negotiations. Although a transgovernmental coalition among European civil servants and transnational pressure from European industry representatives played an important role in raising the pressure on the German government, only the increase of domestic support for the reduction of dual-use controls among the coalition parties, the relevant ministers and their departments, ensured the reduction of national controls by the government. Indeed, the case study shows that the final change in German dual-use control policy only came about when a majority of both national and international actors supported the relaxation of the German dual-use export controls. Conversely, the common European dual-use transfer agreement which was reached ten days later, reflected the varied preferences of Germany, its EC partners and EU institutions by permitting national exceptions in a range of sensitive areas.

Notes

1. The EC was renamed European Union in 1995.
2. P_1 [EU-Co] = 23/28 = 82%.
3. The staff in the Economics Ministry was under pressure from 12 out of 31 actors to whom they had regular contacts in the network, i.e. the representatives of the German industry, their colleagues in France, Britain, Italy, Spain, the Netherlands, Belgium, Portugal, Denmark, Greece, Ireland, Luxembourg and the European Council of Ministers, equalling P_1 [Em] = 12/31 = 39%.
4. With the representatives of the German industry and their colleagues in France, Britain, Italy, the Netherlands, Belgium, Portugal, Denmark, Greece and Luxembourg, 11 out of 25 actors to whom the Defence Ministry staff were linked, i.e. P_1 [Dm] = 10/25 = 44%, called for less restrictive controls.
5. Economics Minister Möllemann was subject to pressure from 12 out of 35 actors including the German industry, his European colleagues and the EC Council of Ministers, with P_1 [EM] = 12/35 = 34%. Whereas Defence Minister Rühe was under similar pressure from his European colleagues who accounted for 10 out of 32 actors to whom he was linked in the network, i.e. P_1 [DM] = 10/32 = 31%.
6. Industry representatives accounted for one of 21 linkages of the CDU/CSU and one of 19 linkages of the FDP parliamentarians, i.e. P_2 [cdu] = 1/21 = 5% and P_2 [fdp] = 1/19 = 5%.
7. Due to the preference change among CDU/CSU and FDP parliamentarians, the pressure on officials in the Defence Ministry increased by two out of 26 actors to which the officials were linked in the network, i.e. from P_2 [Dm] = 10/25 = 40% to P_3 [Dm] = 12/25 = 48%.
8. Similarly, the pressure on officials in the Economics Ministry increased by two from P_2 [Em] = 12/31 = 39% to P_3 [Em] = 14/31 = 45%.
9. That is P_3 [Dm] = 12/25 = 48% increased to P_4 [Dm] = 13/25 = 52% and P_3 [Fm] = 13/33 = 39% to P_3 [Fm] = 14/33 = 42%.
10. That is P_3 [Dm] = 12/25 = 48% increased to P_4 [Dm] = 13/25 = 52% and P_3 [Fm] = 13/33 = 39% to P_3 [Fm] = 14/33 = 42%.
11. P_4 [FM] = 13/38 = 34%.
12. P_5 [DM] = 12/32 = 38%.
13. P_6 [FM] = 16/38 = 42%.
14. Specifically, seven out of 15 actors to which the Coalition Meeting was linked in the network, namely Economics Minister Rexrodt, Defence Minister Rühe, officials from the Economics Ministry, the Defence Ministry, CDU/CSU and FDP parliamentarians and industry representatives, raised the pressure to P_6 [Coa] = 7/15 = 47%.
15. The Cabinet was under pressure from Economics Minister Rexrodt, Defence Minister Rühe, the Chancellor's Office minister, officials from the Economics Ministry, the members of the CDU/CSU and FDP coalition parties as well as industry representatives, who accounted for seven out of its 20 links in the network, i.e. P_6 [Cab] = 7/20 = 35%.
16. Carol Reed and Heinz Schulte, 'Germany views the way to go', *Jane's Defence Weekly* 21:24, 18 June 1994, pp.52f., p. 52.
17. P_7 [FM] = 16/38 = 42%.
18. P_7 [Coa] = 8/15 = 53%.
19. P_7 [Cab] = 8/20 = 40% to P_8 [Cab] = 9/20 = 45%.
20. P_8 [Cha] = 19/42 = 45% and P_8 [FM] = 17/38 = 45%.
21. Frequency = 'changed' / 'changed'+'unclear or undecided'+'no change'
    Instances of blocked preferences are excluded from the calculation because they are explained by the second hypothesis.

4   Transatlantic Community:
Air Strikes in Bosnia

Introduction

As has been argued in the introduction of this book, European foreign policy cannot be reduced to the EU. Most European governments are members in a number of other international organizations, such as NATO, the OSCE and the United Nations, which continue to play a central role in the making and implementation of European foreign policies. Moreover, these organizations link Europe closely with the United States and Canada.

This chapter investigates the utility of multilevel network theory for explaining how these international organizations and the relationship with the United States influence European foreign policy making and vice versa. Specifically, it examines the transatlantic community's endorsement of air strikes in Bosnia in 1993. Although the breakup of the former Yugoslavia was primarily perceived as a European crisis, it soon became apparent that an international response to the conflict went beyond the capabilities of the EU. As a consequence not only the UN Security Council and NATO were called in to deal with the conflict, but also the United States. In order to understand the relationship between these two sets of actors and the making of European foreign policy, this case study focusses in particular on the example of the United Kingdom which had been one of the most ardent opponents of air strikes.

The question whether the international community should use air strikes in response to the Serb attack on Bosnia-Herzegovina[1] arose formally with the UN Security Council resolution 752 which was approved on 15 May 1992. In the resolution, the Security Council condemned the invasion and demanded the withdrawal of Serb forces from Bosnian territory. The Security Council resolution acknowledged that the international community had a responsibility to seek to bring an end to the fighting. In order to do so the UN Secretary General was tasked with an examination of ways to protect human aid deliveries to the Bosnian capital Sarajevo and initiate an international peacekeeping operation. It was agreed that additional measures would be considered in the light of the further development of the crisis.

During the following year, the international community increasingly engaged in various efforts to contain and end the conflict. The main policy options had already been considered after the Serb invasion of Croatia in 1991. They included international negotiations, economic and military sanctions, humanitarian aid and military intervention. Although peaceful measures had been of little success in Croatia, the members of the UN Security Council initially ruled out military action in Bosnia. In spite of the danger that large numbers of refugees might flood into Northern Europe, the five permanent members agreed that the conflict did not affect any of their vital security interests as to justify such a step. Nevertheless, over the course of the year the positions of the Security Council members with regard to military action changed radically. Following foreign policy changes in the US and Britain, the UN Security Council decided on 4 June 1993 to authorize air strikes to protect the safe havens which it had established around the cities of Sarajevo, Tuzla, Zepa, Gorazde, Bihac and Srebrenica. One week later, the North Atlantic Council endorsed the decision.

Two factors mean that an analysis of the American and British contributions to the UN Security Council decision to authorize air strikes in Bosnia is especially interesting for the testing of the hypotheses put forward by multilevel network theory. First, between the recognition of the Bosnian crisis by the UN in May 1992 and UN resolution 836 in June 1993 which authorized air strikes, first the US and, half a year later, the British government effectively reversed their official foreign policy on the issue of military action. According to multilevel network theory, these reversals should have been matched by the strategic interactions in the multilevel European foreign policy network and the coming about and sequence of changes in the preferences of key actors. The decision of the two governments, the UN Security Council and NATO to approve the authorization of air strikes should have coincided with the formation of a winning coalition in favour of air strikes among the actors to which each of these decision-making units was linked.

Second, as in the case of dual-use export controls, exogenous factors were relatively stable and cannot fully account for endorsement of air strikes. The fighting between the Serbs and the Bosnian Muslims continued irrespectively of the various attempts to solve the crisis which were coordinated by the CSCE, the EC, the UN and their negotiators Lord Carrington, later Cyrus Vance and Lord Owen. The cease-fires and peace settlements, which were achieved by the negotiations, were too short-lived to have brought about more than a temporary change in the perceptions and preferences of the network actors (Moodie, 1995: 106; Ware and Watson, 1992: 2). The observed variances in the foreign policy preferences of actors in the network should, therefore, be explicable on the basis of the interactions among these actors themselves.

In particular, the notion that the air strike resolution in June 1993 represented the end point of a rational escalation of means is put into question. Thus, the following analysis demonstrates that air strikes were not only considered from the beginning of the conflict, but also that most of the military experts had believed that an early and massive use of military force could have prevented the escalation of the war. In fact, the more the UN engaged in a humanitarian operation, the greater became the obstacles and risks associated with air strikes. The governments of Britain and France, the two countries which had the largest contingents of peacekeepers in Bosnia, naturally had strong inhibitions against military action. They feared that the Serbs would retaliate and attack their ground troops if air strikes were implemented. Their fears were vindicated in 1994 when the Bosnian Serbs responded to military action in defence of the no-fly zone by taking UN peacekeepers hostage. Air strikes were not an escalation of the international involvement in Bosnia, but an alternative policy which conflicted with the logic of the ongoing humanitarian operation.

Air Strikes in Bosnia: From Opposition to Acquiescence

The conflict between the Serbs and other ethnic groups in Yugoslavia started as early as the beginning of 1990 with human rights abuses in Kosovo and the rise of the autonomy movement among the local Albanian minority. However, the breakup of the Yugoslav republic only progressed from summer 1991 (Gow, 1997; Ramet, 1996). When on 25 June Slovenia and Croatia formally declared their independence, Serb forces moved into Slovenia to prevent its secession. The key members of the UN Security Council and the North Atlantic Alliance decided at this stage that their vital security interests were not at stake (Gow, 1997: 206; Jakobsen, 1994: 5; Macleod, 1997: 250). The concern among Western governments was limited to keeping the conflict at bay and civilians within the warring republics. To achieve its aims, the international community embarked upon the long-term effort to contain the conflict by economic and military sanctions. In addition, it made an attempt to resolve the conflict through negotiations under the auspices of the EC and the CSCE.

After the fighting in Croatia subsided in spring 1992, the conflict moved to Bosnia (Xhudo, 1996). Although the UN had already despatched a small peacekeeping mission to Croatia, the spread of the conflict to another part of the Balkans again presented the international community with the question of the terms of its involvement (Dodd, 1995: 9; Halverson, 1996: 9).

Bosnia provided an opportunity to reassess previous policies and, if deemed necessary, to modify them. The aims and options were essentially the same. Yet, in spring 1992, they had already been debated once and their implications were known.

The option of seeking a peaceful settlement through international negotiations under the auspices of the CSCE, the EC and the UN had so far proved unsuccessful in solving the conflict. The efforts of the CSCE had been inhibited by its recent reform and new, untested, procedures, such as the crisis mechanism. Moreover, lacking sanctions, the CSCE relied on the cooperation of the warring parties - a condition for which there was little prospect in 1991. The consensus rule in the CSCE provided the Yugoslav government and, in their support, Russia with the opportunity to veto stronger action. Similarly, the EC was neither prepared nor suited for negotiating a peace in the former Yugoslavia. The end of the fighting in Slovenia and Croatia was marked by the EU-negotiated 'Brioni Accords'. However, it was the achievement of Serb objectives in Croatia, namely the control over the Krajina, and the reassessment of their interests and their ability to force Slovenia to return to a unified Yugoslavia, that had induced the Serb leadership to accept the cease-fire. The disunity among the EC member states, which was displayed over the issue of the recognition of Slovenia and Croatia, partly accounted for the EC's inability to contain Serb expansion (Macleod, 1997: 245; Gow, 1997: 54). Some member states seemed, albeit unintentionally, to encourage Serb aggression by proclaiming their support for a united Yugoslavia. Others threatened to deploy an international peace force which strengthened the secessionist movement. In short, the European governments' messages were far from clear.

Embargos also had only a limited impact. The economic sanctions which had immediately been imposed by the EC could only be effective in the long term. Yet, when the Serb economy eventually showed signs of weakening, it did not appear to affect military capabilities or the willingness of the Serb leadership to continue the war. The weapons embargo, which had been initiated by the UN in September 1991, had been equally futile. After a delay of the maritime control of the embargo by the WEU and NATO until July 1992, it took another four months to authorize the patrolling ships to pursue and search suspected offenders. Generally, the best that could have been expected from the embargo was a freeze of the military capabilities in the former Yugoslavia. But any hopes that the embargos would lead to an end of the fighting ignored the fact that all republics, especially Serbia, had built up considerable weapons arsenals over the preceding years (Ramet, 1994: 201).

The third policy option, the UN peacekeeping force, had been successful in providing humanitarian relief. Moreover, it had helped to alleviate the consequences of the war for the international community. In particular, it had contained the exodus of the civilian population. However, it could not resolve the conflict.

Due to the apparent lack of success of peaceful measures and the repeated disregard shown by the warring parties to their own commitments (Xhudo, 1996: 85), the fourth option - the use of military force in the form of ground troops or air strikes - had been discussed as early as 1991 (Gow, 1997: 160; Halverson, 1996: 5). Since Western governments were unanimous in their resolve not to become embroiled in a ground war, military action soon became synonymous with air strikes. The possible objectives of air strikes were fourfold: reducing (Serb) military capabilities, safeguarding human aid, pressurizing the warring factions into negotiations and imposing a peace settlement by force. The military capabilities required for the operation could be provided by NATO or its member states in an *ad hoc* arrangement similar to the Gulf War 'Desert Storm'. While the international community agreed that the costs of imposing a peace settlement in the former Yugoslavia was out of proportion with their interests in the region, the first three options matched the need for action with the costs which governments were willing to consider. Nevertheless, the following analysis of the decision-making process from May 1992 to April 1993 illustrates how it took nearly a year for the advocates of air strikes within the transatlantic community to gather sufficient support for a military intervention.

Air Strike Advocates Reach Early Impasse

At the beginning of the debate over air strikes in mid-May 1992, the British opposition to military action reflected the distribution of preferences among key domestic actors within the European foreign policy network which stretched from Britain to the European and North American governments, their foreign and defence ministries and the international organizations involved in foreign and security matters, such as the EU, the WEU, NATO, the OSCE and the UN. While the pressure for air strikes was too low to elicit organized opposition, the overwhelming majority of British foreign policy actors rejected military intervention as disproportionate and unsuitable. In fact, the salience of the issue was so low that some actors within the network, such as the British Labour Party, did not have a public position on the question of air strikes at the beginning of summer 1992.

The British government in particular was strongly opposed to military intervention. Although Prime Minister John Major was careful not to exclude military strikes as a last resort, the extensive conditions set by his government for such action *de facto* ruled air strikes out (Gow, 1997: 95). Indeed, at the beginning of the summer, the British government was unwilling to consider any form of active participation in Bosnia, including the provision of ground troops to an international peacekeeping force. Foreign Secretary Douglas Hurd and Defence Secretary Malcolm Rifkind concurred with their staff's assessment that the war in the former Yugoslavia did not affect British interests such as to justify the risks involved in military action. Officials in the Ministry of Defence (MoD) were also strongly opposed to air strikes. The military argued that in the absence of a clear political imperative a half-hearted intervention would lead to disaster, as the American involvement in Vietnam had shown before. Conversely, in the British Foreign Office opinions had initially been split. After Serb forces had invaded Croatia at the end of 1991, a small number of senior officials had considered the threat of air strikes. However, by May 1992 the internal consensus in the Foreign Office was that military intervention was not politically viable. Part of this conclusion was due to the fact that the British public and members of Parliament (MPs) had so far taken little notice of the conflict in the former Yugoslavia. The country was of little direct relevance to the United Kingdom and lacked a strong national lobby.

In Britain, Lady Margaret Thatcher was the only prominent political figure in the government party who advocated air strikes. However, her calls for military intervention were met with incomprehension, if not embarrassment, within the Conservative Party. Conservative MPs fully backed the position of the government, as did the Labour opposition members when they eventually took a stance on the issue at the end of the summer. Essentially Labour MPs shared the government's assessment that the conflict did not justify action which would put the lives of British soldiers in jeopardy.

Among the British actors in the multilevel European foreign policy network, the Liberal Democrats were the only outspoken proponents of air strikes at the beginning of summer 1992. However, their ability to influence the policy of the government was restricted by their weak position in the European foreign policy network and the absence of further supporters of military strikes in Britain. In particular, Liberal MPs lacked direct influence over the administration since their linkages with government ministers and civil servants in the key departments were characterized by dependence. The Liberal MPs' only options for exerting pressure on the government were indirectly through Parliament and the Parliamentary Defence Committee, or by using their relations with the electorate and the media. However, within the first two, the position of the Liberal Democrats was very weak. Due to the first-past-the-post system, the party had won only 20 out of 651 seats in the previous general election. Proportionally, the pressure of the Liberal MPs amounted to only 9 per cent within the Commons.[2] In the House of Commons Standing Defence

Committee, where the Liberal Democrats were granted a single seat, their influence was negligible.

The highest probability of affecting the preference of another actor lay with representatives of the media who were exposed to much higher pressure than Parliament as a collective decision unit. While the relations of the House of Commons were mainly restricted to the government and the parties represented in it, the journalists were dependent on information not only from domestic, but also international actors. Accessible to pressure for military intervention from a wide range of actors, the international press was used specifically by the governments of Germany, Italy, Turkey, Portugal, the Netherlands, Belgium and Austria to promote the option of air strikes. In fact, together with the Liberal MPs, politicians and officials from these seven administrations accounted for 20 per cent of the actors from whom journalists gained their information.[3] Since this pressure was higher than on any other actors to which Liberal MPs were directly linked, party leader Paddy Ashdown acted rationally in his use of network relations when he concentrated his efforts on a media campaign to raise the prominence of the conflict in Bosnia, rather than attempting to exert pressure on the government via Parliament. In particular, Ashdown travelled repeatedly to the former Yugoslavia, trailing groups of journalists who reported directly back from the war-torn country to the British public. In addition, Ashdown published several articles in the *Guardian* in which he urged for international intervention in Bosnia. The Liberal Party leader also approached the government in personal letters and was eventually granted private meetings with Prime Minister Major and Foreign Secretary Hurd. Although Ashdown was unable to influence the policy of the government directly, the meetings furthered his cause by receiving considerable media attention.

Continental Europeans Favour Air Strikes

Between May and August 1992, the strongest advocates of air strikes were politicians and officials from Germany, Italy, Turkey, Portugal, the Netherlands, Belgium and Austria. Unlike members of the Liberal Democratic Party, these governments had a considerable number of influential relations with key actors in the British government and international organizations. Since their ministers and officials were in regular bilateral and multilateral contact, not only with each other, but also with their British counterparts, they were particularly able to press the British administration on the air strike question. In fact, collectively the foreign and defence ministers from the seven countries accounted for 22 per cent of the linkages of Foreign Secretary Douglas Hurd in the network. Their pressure was only marginally lower on

Secretary of Defence Malcolm Rifkind for whom they represented 21 per cent of his regular contacts.[4] Practically, it meant that nearly a quarter of the actors to which the two British ministers were directly linked within the multilevel European foreign policy network urged them to implement air strikes in Bosnia. In the case of Foreign Office and MoD officials, the degree of pressure was even higher at 27 per cent[5] and 29 per cent[6] respectively because they had fewer linkages than their ministers, and their colleagues in Germany, Italy, Turkey, Portugal, the Netherlands, Belgium and Austria represented a larger proportion of their relations.

However, although the staff in the Foreign Office and the MoD were under acute transnational pressure, the strong consensus within the British executive and Parliament helped them to resist the calls for air strikes. Most of the domestic actors with whom civil servants from the two ministries had regular contacts, such as their ministers, the Cabinet, Prime Minister Major and officials from other ministries, maintained their opposition against military strikes. While significant, the pressure on the officials from the Foreign Office or the MoD was not sufficient for them to abandon their doubts. Crucially for the decision-making process, the resistance from the officials in the Foreign Office and the MoD prevented the formation of a transnational coalition among the bureaucracies in favour of military action during summer 1992. Their boundary position, thus, enabled them to insulate a large number of British actors in the network from international pressure.

International Organizations Under Pressure

However, ministers and officials from the European governments who favoured air strikes had alternative network relations which they could use to persuade or press other actors to support air strikes. Specifically, they were able to use their common membership in the international organizations which had become involved in the crisis, i.e. the EC, NATO, the WEU and the CSCE, to form an intergovernmental coalition. The rationale behind this strategy was that the participation of these organizations, in the resolution of the conflict in the former Yugoslavia, would provide them with indirect influence over the British administration if the organizations adopted a favourable stance towards air strikes. Moreover, the nature of these organizations as collective decision units ensured that the preferences of all members were reflected. Due to differences in the membership of each organization, the balance of pressure for military action was highest within the WEU where 53 per cent of the actors linked to it favoured air strikes.[7] The pressure was marginally lower in the NATO's integrated military organization at 45 per cent,[8] the European Council of Ministers at 40 per cent,[9] the North Atlantic Council at 37 per cent[10] and the

CSCE at 35 per cent.[11] The UN Security Council was almost free from direct pressure for air strikes since its permanent members, in particular the US, Britain, France and Russia, were opposed to military action. Latent support for air strikes from US Secretary of State James Baker raised the pressure to a mere 10 per cent.[12]

In spite of the considerable pressure in these international organizations, their limited authority over an international operation like air strikes impeded the effectiveness of the international coalition in favour of military action. The CSCE was still in the process of being transformed from a series of conferences into an international security organization. Although several mechanisms had just been established which permitted CSCE delegations to monitor and intervene peacefully into the solution of conflicts among its member states, military action was beyond their capabilities. Similarly, the European Council of Ministers had neither the military means nor the legitimacy to order air strikes in Bosnia. Recent attempts to include the WEU into the structure of the EC in order to provide it with a defence arm had not been very successful. The strengthening of the EC's role in foreign and security policy matters had been prevented by the disagreement among member states over the degree to which an EC-WEU force should be independent from NATO. As a result, the Franco-German Euro-Corps remained the only integrated military unit under the command of the WEU. However, as a lightly equipped land force the Euro-Corps was not more suitable for a full-scale intervention than for selective air strikes. Moreover, the corps was not yet operational. In order to implement air strikes, the EC Council of Ministers would have to call upon the member states of NATO or the WEU to act on its behalf. This would mean transferring the decision-making authority over the international involvement in Bosnia to these organizations. However, in terms of international law, an intervention by NATO, the WEU, or an *ad-hoc* arrangement between the major powers comparable to that in the Gulf War could only be authorized by the UN Security Council. This would remove the issue further from the influence of the European advocates of air strikes. Moreover, NATO and the WEU were not yet legitimized to take action beyond the territory of their member states. Since the treaties of both organizations stipulated their functions as collective self-defence rather than collective security, the intervention into a conflict that did not involve any of their member states was outside their scope.

Vetos Prevent Action

While the functional scope of NATO and the WEU could be changed, just as the CSCE had taken on new tasks, the most serious obstacle lay in the decision-making structures of the two institutions. Both provided member states with a veto due to their formal requirement of a consensus for all Council decisions. While multilevel network theory suggests that a veto can be overruled if the pressure is sufficiently high, it also contends that a veto of one or several of their members allows organizations, such as NATO and the WEU, to withstand higher degrees of pressure than actors without veto rules such as in ministerial departments or political parties. Indeed, although the pressure for air strikes was considerable in all organizations, their veto position not only allowed member states which opposed military action to block an endorsement of air strikes, but also the extension of the authority of these organizations to implement them.

In the CSCE, the Russian and Serbian veto effectively prevented a direct involvement of the organization in most of the international operation in Bosnia. While uncontroversial tasks, such as fact-finding missions, had been permitted at the beginning of the conflict, by 1992 it had been recognized that due to its large membership and the prominent Russian veto of outside intervention, the ability of the CSCE to end the war was restricted. The EC Council of Ministers was equally divided over military action, in spite of the smaller number of its member states and their greater homogeneity. In fact, the failure of the EC to achieve any significant progress in the peace negotiations which were under its authority had been partly attributed to the conflicting signals emanating from its members' governments (Gow, 1997: 54). The split within the EC had its origins in the debate over the recognition of Croatia and Slovenia in 1991 (Macleod, 1997: 245).

UN Takes Authority Over International Response

The failure of the EC Council of Ministers to agree on measures other than economic sanctions, while the UN Security Council decided upon and implemented a United Nations Protection Force (UNPROFOR) to safeguard aid deliveries in Bosnia, eventually contributed to the transfer of the issue to the primary authority to the Security Council. In fact, some member states appeared to block action by the EC, NATO and the WEU with the aim to pass the responsibility over the international response to the Yugoslav crisis to the Security Council. The French government in particular was wary of its lack of influence over NATO's military decisions if air strikes were considered. Its representatives, therefore, vehemently demanded a UN resolution to back such action.

The French interposition came at a time when the supporters of air strikes began to focus their pressure on enlarging the functional scope of NATO in order to prepare it for intervention out-of-area, such as in the former

Yugoslavia. At the beginning of June, the Dutch Defence Minister took the lead on the issue by proposing to make NATO forces available to UN and CSCE peacekeeping operations outside the borders of its member states. Recognized as a necessary first step towards military action in Bosnia, the proposal was supported by the representatives of Germany, Italy and Turkey. The governments in Britain and the US were divided over the issue. On the one hand, both administrations had long been in favour of enlarging the functional scope of NATO. It would enable them to more easily use NATO's integrated military structure in cases like the Gulf War. On the other hand, most British and American politicians and officials were strongly opposed to similar involvement in the former Yugoslavia. Doubts over the transformation of NATO were also expressed by the governments of Spain and Belgium, in spite of the latter's support for air strikes in Bosnia. The strongest opposition came from the French administration which not only disapproved of military strikes in Bosnia, but also of the extension of the functions and the authority of NATO. With regard to military action in Bosnia, the French government preferred the authority to rest with the UN Security Council where it had a veto and where the consensus was against air strikes. As to out-of-area missions, the French administration envisaged a global peacekeeping role for the WEU (Lepick, 1996: 79). Since the representatives of Britain and the US sided with the advocates of air strikes on the question of out-of-area missions, the French veto was easily overruled in the North Atlantic Council.

Nevertheless, on the issue of Bosnia, the French government succeeded with the support of the British and Americans in blocking further moves towards military action in the North Atlantic Council. The representatives of the three governments insisted that such measures could only be decided on the basis of a UN Security Council resolution. The demand implied the transfer of the ultimate decision-making authority over the international operation in Bosnia to the Security Council. Although the transfer of the authority was not favoured by the air strike advocates, they were not able to prevent it. The EC and NATO had proved incapable of agreeing on effective measures. Moreover, the support of the three Security Council members, France, Britain and the US, was in any case a condition for air strikes because of their military contributions to NATO or the WEU. In the absence of any prospects for an agreement, the transfer at least relieved the other European governments from their responsibility with regard to the conflict. Essentially, however, the coalition between the governments of Germany, Italy, Turkey, Portugal, the Netherlands and Austria had reached an impasse. Although their ministers and officials continued to press for air strikes in the EC and the WEU, the sidelining of these organizations on the Bosnia issue deprived them of direct control over the international intervention. Since the advocates of air strikes had used all their linkages within the network to exert pressure, but not succeeded in gaining new supporters, the option of air strikes appeared to have been finally ruled out. However, at the same time the pressure for air strikes was reaching critical degrees in the US. The new Secretary of State Lawrence Eagleburger and President George Bush were especially susceptible to further pressure when news about Serb ethnic cleansing in Bosnia emerged at the beginning of August 1992.

Concentration Camp Report Triggers US Policy Change

The next stage of the debate over air strikes in Bosnia was marked by a decisive shift of US foreign policy towards military action. It was followed by attempts from US officials to press their French and British partners in the UN Security Council into supporting air strikes. The American policy change in favour of air strikes was triggered by a broadcast from Independent Television News which showed 80 starving prisoners in a Serb concentration camp near Omarska. As the pictures flashed about television screens in the US and Europe, voters and parliamentarians who so far had been largely ignorant about the conflict were outraged (Halverson, 1996: 8; Ramet, 1994: 204). Within days leading Congressmen and Senators urged the US administration to intervene militarily in order to end the atrocities. Although Congress did not have any immediate decision-making authority in this case, its direct influence over the President and the Secretary of State due to the general institutional dependence of the US administration on Congress in other matters allowed its members to exert considerable pressure. Journalists through their own relations as provider of information for the government and as mediator of the public opinion reinforced the demands of Congress members by reports about other Serb camps in Bosnia and commentaries on the viability of military action.

As opinions changed, they profited from the latent support for air strikes within the State Department. Officials within the State Department had for some time been dissatisfied with the government's reluctance to take action in Bosnia. In fact, several had resigned in protest of the official policy (Gow, 1997: 211; Ramet, 1994: 204). When the ITN broadcast finally raised the public awareness of the conflict in the former Yugoslavia, members of the State Department authenticated the news about the killing of civilians in the Serb camps without consulting their political leadership. Although Assistant Secretary of State Thomas Niles was subsequently forced to deny the announcement, his assurance that the administration could not confirm reports of Serb death camps now lacked credibility.

President Bush was particularly exposed to the rising pressure from members of Congress, the media and the American public. Not only was he linked, through a web of formal and informal relations to Senators and journalists, Bush also had to be especially wary about the view of the American electorate due to the upcoming presidential elections (Halverson, 1996: 10). The resolve of the Democratic candidate Bill Clinton, who called for selective air strikes to end the conflict, appeared more favourably in the American public than the wavering of President Bush. Within two weeks the national and international pressure on the President increased from 28 to 34 per cent of his contacts in the network[13] and Bush began to sway. Although the Secretary of Defence Richard Cheney and officials from the Pentagon reiterated their opposition against military action, Bush indicated an impending change of policy with regard to the military strikes on 16 August 1992.

The Secretary of Defence and his staff, however, resisted the calls for air strikes from 42 and 37 per cent of the actors with whom they had contacts in the network.[14] They were supported by their colleagues in NATO who agreed on the dangers of intervening in Bosnia. Crucially, the continuing internal disputes between the 'hawkish' officials in the State Department and the 'dovish' military in the Pentagon seriously reduced the ability of the administration to press for military action (Watson and Ware, 1993: 7). In fact, the officials from both departments used their relations within the network to pursue their divergent policy preferences, sometimes undermining the position of the US government in international organizations and among its European partners. While State Department officials embarked upon a policy of persuasion and negotiation with America's closest allies in the UN Security Council during autumn in order to promote the new US policy (Jakobsen, 1994: 20), Pentagon staff used their linkages within the US administration and NATO's integrated military organization to urge caution.

Pressure on Britain and France Rises

Since the transfer of the ultimate decision-making authority to the Security Council, the main targets for pressure from State Department officials had to be their colleagues in Britain and France who continued to veto air strikes. Although the European press and public were equally outraged at the ITN pictures of the Serb detention camps as the Americans, they shared the scepticism of their political representatives as to whether military strikes could resolve the conflict in Bosnia. Risking the lives of British or French military personnel without a clear objective or strategy appeared self-defeating (Gow, 1997: 211). Nevertheless, the calls for military action had increased in Europe as well. Among the new proponents of military action were the members of the European Commission, represented by Jacques Delors and the External Relations Commissioner Hans van den Broek and the European Parliament.

The Commissioner who had so far been neutral on the issue responded to calls for air strikes from the German, Italian, Portugese, Dutch and Belgian administration, which together with the media accounted for 41 per cent of the Commission's links within the network.[15] Moreover, without immediate authority or responsibility over the international operation in Bosnia, the Commission could safely take a radical stance. Speaking to the European Parliament, Commission President Jacques Delors advocated air strikes as the new solution to the Yugoslav quagmire. Among the European Parliament members his views were increasingly shared. However, in spite of the growing pressure from EU institutions, the representatives of Britain and France continued to veto the endorsement of air strikes in the EC Council of Ministers.

While President Bush was still wavering about his position on air strikes during the second week of August, officials from the US State Department began to focus their pressure for military action on diplomats in the UN Security Council. Specifically, State Department officials proposed a resolution authorising the use of 'all necessary means' in Bosnia. Although the phrase had obvious similarities with the UN resolution which preceded the international intervention in the Gulf War, the formulation was sufficiently vague to accommodate the diverse positions in the US, Britain and France. It stipulated neither the form of the measures to be taken, nor a deadline for an intervention. The British and French administrations accepted resolution 770 to placate the media and the increasing proportion of their public who were favouring military action.

No Air Strikes Despite UN Resolution

The British and French acquiescence to a resolution which could pave the way for air strikes gave an indication of the increasing pressure on the two governments with the emerging shift in US foreign policy. However, accounting for 23 per cent, the number of air strikes supporters among the actors to whom the British Prime Minister John Major was linked in the network, was still significantly lower than the 34 per cent which led President Bush to adopt air strikes as an official policy objective only days after the UN resolution.[16] Nevertheless, the resolution effectively committed the governments in Britain and France to some form of action in Bosnia. The French administration was certainly aware of the implicit change of policy. The question which international organization would implement military action, if it could be agreed upon, already caused disagreements between US and French officials. The latter insisted that the UN should maintain the command over the operations in Bosnia, including military action implemented by NATO. Conversely, the US administration preferred a fully independent NATO mission. The outcome of this debate was a nominal compromise which revealed how far the Security Council was from considering military action in practice. In the text of resolution 770, which was approved on 13 August, the members of the Security Council declared to 'take nationally or through regional agencies or arrangements all measures necessary to facilitate in coordination with the United Nations' the delivery of aid to Bosnia. Practically, it meant that the specific nature of the intervention would be decided upon when the UN Security Council members had reached a consensus on the question.

If US State Department officials had assumed that resolution 770 would open the way for air strikes in Bosnia, their hopes were soon shattered. Not only did their colleagues in Britain and France insist that all other measures were tried before military action was even considered, the American plans for air strikes also met the resistance of a Pentagon-led transgovernmental coalition among the military staff in NATO. The military used the fact that the UN Security Council had tasked NATO with examining the options for the implementation of resolution 770 to question a military intervention in Bosnia. The suggestion to seize the Bosnian capital Sarajevo with a 100,000 strength force made by military officials from the headquarters of the Atlantic Alliance in Brussels was, as newspapers commented, 'an impractical proposal designed to get a thumbs-down from Congress, the Bush administration and NATO'.[17] It was well known to military officials that neither American, nor European politicians were willing to despatch fighting troops to the former Yugoslavia. However, NATO commanders claimed that this was necessary if the international community wanted to enforce UN resolution 770 by military means.

In addition, Pentagon officials and their British colleagues used their domestic relations in the multilevel European foreign policy network to repeat the estimates in conversations with their political leaders and with parliamentary representatives. In the US, the assistant to General Colin Powell, Lt General Barry Mc Caffrey, told the members of the Senate Armed Services Committee that between 60,000 and 120,000 troops were required to secure Sarajevo for the delivery of humanitarian aid as envisaged by the UN resolution. He pointed out that such military intervention would certainly involve a high number of casualties. When questioned by the British government, the Chiefs of Staff even topped the dire scenario painted by the American military. They suggested the creation of corridors for aid convoys to Sarajevo would require at least 300,000 fighting troops.

Britain and France Send Ground Troops

As both European and American politicians were alarmed by the vision of becoming embroiled in the Yugoslav war, the prospect of a military solution again received a setback. It was hardly noted that the less risky option of air strikes on Serb bases had been *prima facie* excluded by the military under the pretence that nothing short of a full-scale intervention would be 'effective'. Thus, with the explicit support from NATO military planners, British and French representatives in the North Atlantic Council were able to maintain their veto regarding air strikes in spite of the increase in pressure from 39 to 41 per cent following President Bush's preference change in support of air strikes.[18]

After NATO military action had been ruled out, the governments of Britain and France decided unilaterally to implement resolution 770 by means of a small, defensive protection force of about 3,000 soldiers. In Britain where the news of Serb concentration camps had not changed the doubts of politicians and the public over military action, the decision was met by broad approval within the executive and legislative. In the House of Commons, even the members of the Labour Shadow Cabinet supported the government's policy.

Effectively the emerging transatlantic coalition in favour of air strikes stalled at the borders of the United Kingdom in October 1992. The effect of the American policy change on actors in Britain was limited. Officials in the British MoD who had been under the highest pressure within the British administration at 29 per cent, were not exposed to American pressure for military action because they were only linked to their colleagues in the Pentagon who continued to oppose air strikes. The main change in pressure was noticed by Prime Minister Major who was now confronted with President Bush's support for military strikes in international summits. While the pressure on the key roles within the British government ranged between 21 and 29 per cent, it did not lead to an immediate response, with its long-term consequences showing during the winter of 1992-93.

US Urge For No-Fly Zone

During autumn 1992 the international pressure on the British administration persisted. Since resolution 770 had failed to induce the governments of Britain and France to shift towards the implementation of military action, US officials embarked on renegotiating the issue in the UN Security Council. The vagueness of the resolution's stipulations had been one of the reasons why the British and the French governments had been able to circumvent military strikes by taking alternative action. With a new resolution, President Bush and

Secretary of State Eagleburger wanted to endorse explicitly the use of offensive means of action. Since the representatives of Britain and France ruled out a straightforward air attack on Serb targets such as airfields and artillery, US State Department officials proposed the extension and subsequent enforcement of the no-fly zone to the Bosnian airspace (Ware and Watson, 1992: 7). The repeated violations by Serb and Croat planes provided suitable justification, although most of the flights had been for logistic purposes and travel, rather than offensive action. In reality, the US were little concerned about the breaches of the no-fly zone. As one official admitted later, the question of pursuing planes in violation of the no-fly zone was seized upon as an opportunity to introduce air strikes on Serb airfields as a logical extension of the enforcement measures. Since the American proposal avoided the mentioning of air strikes, the members of the UN Security Council quickly agreed on the principle of extending the no-fly zone over the former Yugoslavia to Bosnia. On 9 October, the Security Council approved, in resolution 781, the extension of the no-fly zone.

It was little surprising, however, that British and French representatives at the UN vetoed any efforts to enforce the zone by shooting down planes caught in violation. Their ability to resist the pressure for air strikes was enhanced by similar inhibitions within the Canadian, Danish and Spanish administrations. The three countries had contributed ground troops to UNPROFOR and shared French and British fears of retaliation. Moreover, the Russian government had been fundamentally opposed to any intervention in the former Yugoslavia since the beginning of the conflict. Air strikes were unacceptable to the Russian leadership, not only because of close historical ties with the Serbs, but also because it might set a precedent for Western intervention into the former Soviet Union and its sphere of influence.

Opinions Begin to Shift in Britain

Nevertheless, the persistent pressure from Eagleburger and US State Department officials, as well as from the administrations of the seven European countries who had advocated air strikes since May, slowly eroded the opposition to an enforcement of the no-fly zone between October and December 1992. British Foreign Office officials were particularly likely to change their view on the issue because 27 per cent of the actors with whom they regularly cooperated in the network supported air strikes.[19] Moreover, due to their involvement in the UN negotiations about the implementation of the no-fly zone, Foreign Office diplomats were constantly exposed to the international demands for military action. At the beginning of December 1992,

Foreign Office officials speaking to the press expressed a changed preference in favour of air strikes.

The consequences of their preference reversal for the formation of a winning coalition in favour of air strikes were considerable. Specifically, the advocacy of air strikes by Foreign Office officials instantly increased the pressure for military action on Foreign Secretary Douglas Hurd from 25 to 28 per cent of his contacts in the network.[20] Only three days later, Douglas Hurd announced in the media that, according to his opinion, the time had come to consider the use of air strikes in Bosnia. Moreover, the support for air strikes from Foreign Secretary Hurd and Foreign Office officials in turn raised the pressure for, and the credibility of, military strikes among MPs and the members of the Conservative and Labour Parties who looked to these two actors for expertise and information on the issue. Of the actors to which Conservatives and Labour members were directly linked within the European foreign policy network, 25 per cent now called for a military intervention.[21] However, while the Labour opposition joined the coalition in favour of air strikes, the same increase in pressure had less effect on the members of the Conservative Party. Although the number of proponents of military action was increasing among the Conservatives, the majority of MPs maintained their support for the government's policy.

Britain Accepts No-Fly Zone Enforcement

The scope of the preference changes among civil servants and politicians significantly weakened the opposition against air strikes in the European foreign policy network. By mid-December, the support for air strikes by Douglas Hurd and Foreign Office officials had increased the pressure on the Prime Minister and the Secretary of Defence to 28 per cent[22] and on the military and civilian staff in the MoD to 33 per cent of their network contacts.[23] Yet, collectively the members of the Cabinet maintained their objections to military action during December. Moreover, as a consequence of the convention of collective responsibility in the Cabinet, Foreign Secretary Hurd was forced to defend the official policy line nationally in the media and internationally in the ongoing negotiations about the enforcement of the no-fly zone. While the preference changes had stopped short of British acquiescence to air strikes, the split within the British administration weakened its ability and willingness to resist the pressure for the implementation of the no-fly zone by air strikes. In the Cabinet's Overseas Policy and Defence Committee, the differences between Foreign Secretary Hurd and Defence Secretary Rifkind were resolved by a compromise in favour of a UN resolution which authorized the enforcement of the no-fly zone, but ruled out air strikes on ground targets.

In the international negotiations the discussions focussed on the question of how to implement the no-fly zone. A draft resolution tabled by the French government with support from the US, the Netherlands and Turkey set the starting point for extended bargaining between the governments over the means of enforcement. In the North Atlantic Council, the representatives of the US and the other European governments which supported military action secured a blank agreement in which NATO offered its military capabilities to the UN for the implementation of the resolution. However, ministers from twelve out of sixteen member states warned about any action which could put the lives of UNPROFOR soldiers in jeopardy. NATO's final communiqué of 17 December 1992 stressed that the effects of enforcement action on the humanitarian operation would be taken into account.

Nevertheless, in the further course of the international negotiations in the UN, a coalition between the British and French representatives, with the support of the Canadian government, broke up because of the increasing tacit support for military action within the British administration. In a meeting with President Bush at Camp David, Prime Minister Major acceded to the use of air strikes as a last resort and after a 30-day warning period. Once the British government had changed sides on the issue, French and Canadian diplomats were not able to prevent further moves towards a UN Security Council resolution. The text of the resolution was eventually agreed along the lines negotiated between Bush and Major.

New US Administration Reviews Policy

The Security Council vote on the resolution, however, was delayed because of the institution of a new government in the US. Although President-elect Bill Clinton had strongly argued in favour of air strikes during his election campaign, the incoming administration was not prepared to endorse the resolution which had been negotiated by its predecessor without a policy review. Instead, the new administration embarked upon a reassessment of the situation in Bosnia and the policy options at its disposal which put a halt to the resolution until March 1993.

Thus, at the beginning of 1993, the debate appeared to have come to a standstill. For a short period in February, it even seemed as if the new US government had reversed its stance on air strikes because of the opposition from Pentagon officials and the governments of Britain and France. The Vance-Owen Plan which at the time had been accepted by the leaderships of the Croats and the Bosnian Muslims raised expectations in the US and Western Europe that a peace settlement was close. However, Bill Clinton's six-point initiative included the enforcement of the no-fly zone as one of its key measures. When the Serbs continued their policy of ethnic cleansing in spite of the Vance-Owen peace plan, the pressure from State Department officials, Senators and the American public which had existed before the election soon came to bear upon the members of the new administration. Although Secretary of State Warren Christopher adopted a cautious stance, while the new Secretary of Defence Les Aspin became an ardent promoter of air strikes, the overall balance of pressures in the administration remained unchanged. By March, Clinton resumed the course of the previous government in favour of air strikes. To exert pressure on their European colleagues US administrators threatened to unilaterally lift the arms embargo on the Bosnian Muslims and to consider air strikes to stop the war. Moreover, State Department officials took advantage of the increasing domestic pressure on the British government by organizing a series of bilateral meetings with their British counterparts to discuss military action in Bosnia.

British MPs Favour Military Action

In Britain, the additional support for air strikes from Foreign Secretary Hurd, Foreign Office civil servants and Labour MPs that had increased the pressure on Parliament from 15 to 59 per cent began to show its effect.[24] Although the Conservative MPs remained divided over the question of air strikes, the support for military action increased to up to a third of the Conservative Parliamentary Party during the spring. Among the latter were the chairmen of the Commons Committees on Foreign Affairs and Defence Policy, David Howell and Nicholas Bonsor, who advised the government to consider stronger action. Yet, these experts spoke as individual MPs. Collectively, the Conservative Parliamentary Party did not challenge the government but supported its policy with its blocking majority in the Commons and its Committees. Noting the growing support for military action, however, the government sought to exert counter pressure on MPs through extensive briefings of the Standing Committees and the Commons about the dangers of military intervention.

The government could do less to prevent a change of opinion among the British public. The electorate had been exposed to calls for air strikes from the media and other actors since August. In December, the balance appeared to be shifting, but the public remained divided over the viability of military action. By the first week of April a MORI poll confirmed that 60 per cent of the British population were dissatisfied with the government's performance and as many as 64 per cent supported a despatch of British troops to Bosnia to stop the fighting. The change of public opinion in favour of military action increased the pressure on MPs and cabinet ministers. Due to his boundary position in the network Secretary of Defence Malcolm Rifkind was not only exposed to transgovernmental pressure from the governments which called for air strikes, but also to the demands of his colleague Foreign Secretary Hurd. The recent swing in opinion polls increased the percentage of actors who supported air strikes among Rifkind's linkages in the network from 31 to 35.[25] The change of the public mood also raised the pressure for air strikes on Prime Minister Major from 31 to 33 per cent[26] and on the Cabinet from 21 to 29 per cent.[27]

No British Veto On Air Strikes

On 25 April, Secretary of Defence Malcolm Rifkind acceded and expressed his support for air strikes. His preference change in turn raised the stakes for Prime Minister Major from 33 to 36 per cent of his network contacts.[28] Ten days later, John Major announced that the government did not rule out air strikes anymore.

As a policy change in the Cabinet appeared close, the NATO military made a final attempt to prevent air strikes through a transgovernmental coalition. At the core of this counter-coalition was the military staff from the British MoD. Unlike their political leaders, the military were not dependent on support from the public and Parliament and, thus, unaffected by their increasing pressure. Conversely, military officers were able to use their position as provider of authoritative information on defence issues for journalists and politicians to try to convince them of the unsuitability of air strikes. In a coalition with their colleagues in NATO, British military leaders decided to focus their efforts on the media. One day before the crucial meeting of the British Cabinet, the NATO Chiefs of Staff issued a statement in the international press in which they warned collectively of the dangers of military intervention. In the Cabinet, Chancellor Norman Lamont, Home Secretary Kenneth Clarke, Social Security Secretary Peter Lilley and Scottish Secretary Ian Lang also rejected air strikes. However, these ministers had little influence on the decision because most of them were not part of the European foreign policy network. In the end, neither they nor the military were able to prevent the formation of a winning coalition in favour of air strikes in the British Cabinet. The collective pressure from the Prime Minister, the Foreign Secretary, the Defence Secretary, Foreign Office officials and the public which represented 43 per cent of the actors linked to the Cabinet was overwhelming.[29] On 30 April, the British government decided not to veto air strikes in Bosnia if it should come to a Security Council resolution.

UN Resolution 836

Despite its decision not to block a UN resolution calling for air strikes in principle the British government, like the French, remained hesitant to use military action in Bosnia for fear of retaliation against its peacekeeping troops. However, during the first week of May both governments came under increased pressure from the US after President Clinton despatched Secretary of State Christopher on a tour of London, Paris, Berlin, Moscow and the NATO headquarters in Brussels to mobilize support for a lifting of the arms embargo on Bosnian Muslims and the implementation of military strikes. Although Christopher's mission appeared to be of little success at the time, by the end of his tour foreign policies in Europe had begun to shift. In particular, the possibility of air strikes was no longer ruled out by either the British or the French governments if certain conditions were met. Both administrations insisted that the arms embargo would have to remain in place since the provision of weaponry to the Bosnian Muslims would inevitably lead to an escalation of the conflict which would endanger British and French peacekeepers. Furthermore, France demanded that the control over air strikes would have to rest with the UN mission in Bosnia which was led by the French General Morillon, and not with NATO where France had less influence.

When the Bosnian Serb parliament rejected the Vance-Owen plan on 6 May, air strikes were thus emerging as a new policy. The failure of the plan which had been favoured by the Moscow leadership also made it less likely that Russian would veto air strikes in the Security Council. As immediate response to the rejection of the Vance-Owen peace plan and the continuation of the fighting in Bosnia, however, the UN Security Council decided to create safe havens around the Bosnian capital Sarajevo and the towns Tuzla, Zepa, Bihac and Gorazde. Crucially, the resolution which was passed the same day implied the threat of air strikes if Serb artillery continued their attacks on the towns. Indeed, while further action was delayed because of marginal hope that a referendum by the Bosnian Serbs might still rescue the Vance-Owen peace plan, European foreign and defence ministers who were meeting within the context of the WEU on 19 May seized the idea of air protection for the safe havens as a compromise. Such action would limit the scope of US air strikes and bind them to decisions by UN military commanders on the ground. The WEU thus responded to overwhelming pressures in favour of air strikes at 74 per cent and marked the end of the British and French veto which had so far blocked a policy change within the organization.[30] Ten days later the US agreed with Britain, France, Russia and Spain on a draft resolution which endorsed this policy and on 4 June the UN Security Council approved resolution 836 at a pressure of 60 per cent.[31] The resolution authorized 'member states, acting nationally or through regional organizations or arrangements, [to] take ... all necessary measures, through the use of air power in and around the safe areas in the Republic of Bosnia and Herzegovina to support UNPROFOR'. The Ministerial Council in NATO, where the balance of pressures had by now reached 55 per cent, endorsed the resolution a week later.[32]

Assessment of the Hypotheses

In order to evaluate the hypotheses proposed by multilevel network theory, this section summarizes the findings concerning the relationship between pressure and preference changes in the context of the transatlantic community. As in the previous cases study, it uses the frequency of preference changes with rising degrees of pressure, the distribution of pressure in the four behavioural categories (no change of preferences, unclear or undecided preferences, change of preferences and blocked preference changes), the average pressure in each behavioural category and the timing of the preference changes as indicators for the validity of the proposed hypotheses.

The first indicator measures the proportion of preference changes among the observed behavioural responses to network pressure within 5 per cent intervals.[33] The findings of this measure are presented in Figure 4.1. They confirm approximately the first hypothesis which suggests that higher degrees of pressure increase the probability that an actor will change his or her policy preference. Thus, none out of five actors who were subject to very low pressure in the range of 0-15 per cent changed their preferences, whereas one of twenty actors who were subject to higher pressure of 15-20 per cent changed theirs. Indeed, the results displayed in Figure 4.1 demonstrate that the proportion of preference changes increased almost monotonously with rising degrees of pressure.

The second indicator, namely the distribution of the four behavioural categories - no change of preferences, unclear or undecided preferences, change of preferences and blocked preference changes - across different degrees of pressure applies to either the first or the second hypothesis. Specifically, the first hypothesis suggests that the number of instances in which actors maintained their preferences decreases with higher degrees of pressure, while those of actors who changed their preferences increases. In addition, the second hypothesis proposes that the blocking of preference changes in collective decision-making units, like the North Atlantic Council or the UN Security Council, enables these actors to withstand higher degrees of pressure than actors without a veto option.

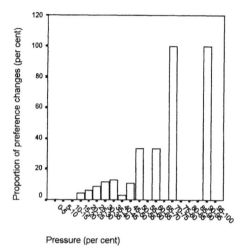

Figure 4.1   Proportion of preference changes

In the case of air strikes in Bosnia, these hypotheses were broadly confirmed. Thus Figure 4.2 shows that most of the actors did not change their policy preferences if the pressure on them ranged between 20 and 45 per cent. However, once the degree of pressure increased beyond 50 per cent not a single one of the actors in the multilevel European foreign policy network was able to hold on to his or her original preference. As in the previous case study, the observation that Figure 4.2 peaks twice at 20-25 and 40-45 per cent pressure can be explained by the choice of the research period which implies that few actors were subject to lower degrees of pressure.

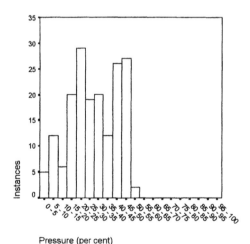

Figure 4.2   No change of preferences

This aside, the distribution of the curve supports the hypothesis that actors are less able to resist demands for changes in their policy preference the higher the pressure from other actors in the network. Moreover, the observation that actors were not able to maintain their opposition to air strikes when the collective pressure on them increased above 50 per cent provides further confirmation for the inductive proposition that there might be a threshold beyond which actors are not able to resist additional pressure. Interestingly, this threshold occurs here at exactly 50 per cent which represents the situation in which half of the actors who have power over a specific member of the network urged him or her to support air strikes.

With regard to the category of unclear or undecided preferences displayed in Figure 4.3, the findings suggest a second inductive proposition. Notably, the finding that no actors were undecided if pressured by less than 15 per cent of the actors to whom they were linked in the European foreign policy network appears to indicate that actors' being unclear or undecided about their foreign policy preferences is also related to the degree of pressure to which they are exposed.

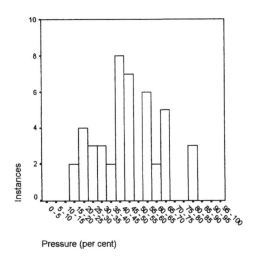

Pressure (per cent)

Figure 4.3   Unclear/undecided preferences

Indeed, the findings appear to suggest a progressive relationship between the three behavioural categories no change, unclear/undecided and change of preferences. Specifically, the observation that the number of actors whose preferences were unclear or who were undecided peaks just before the 50 per cent threshold seems to imply that being unclear or undecided might be an intermediate stage which signifies that those actors are reconsidering their policy preferences. However, the rather high number of instances in which actors were undecided at pressures from 55 to 70 per cent of the actors who had power over them might also indicate that unclear preferences can help actors to withstand pressure for a modification of their policy preferences.

The distribution of preference changes in Figure 4.4, is an even better indicator of the explanatory value of the first hypothesis. It shows the instances in which actors modified their preferences in relation to the pressure to which they were exposed. As hypothesized, the figure shows a steady and almost monotonous increase in the number of preference changes with rising degrees of pressure. Moreover, the evidence indicates that actors only began to change their preferences if the pressure on them increased above 15 per cent. Although preference changes occurred in three instances at pressures above 50 per cent, the figure broadly confirms the inductive proposition of a threshold at a degree of pressure of 50 per cent. Crucially for this assessment, the three exceptions from this proposition which can be found in the figure referred to actors who for much of the research period had been blocked from a preference change, namely the UN Security Council, the WEU and the North Atlantic Council. The ability of these actors to withstand higher degrees of pressure is thus explained by thesecond hypothesis.

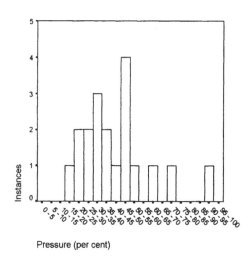

Pressure (per cent)

Figure 4.4 Change of preferences

The validity of the second hypothesis which proposes that a veto or blocking strategy used by one or several actors within a collective decision-making institution allows them to resist higher degrees of pressure than actors without veto rules is indeed corroborated by the evidence in Figure 4.5. Blocking behaviour occurred only if collective decision units were subject to pressure between 35 and 85 per cent. Therefore, the evidence not only indicates that blocked actors could maintain their original policies despite increasing pressure, it also implies that actors did not use a veto or blocking strategy until the pressure on their collective body had reached significant levels.

Moreover, the case study confirmed the proposition that actors might block a decision within one multilateral organization an attempt to transfer the authority over the issue to others. In particular, these actors will try to shift the authority to those institutions which have fewer veto-holding members or within which the balance of preferences is more in their favour. Both strategies are linked in that the inability to agree on a decision within subordinate decision-making units often leads to the transfer of the issue to units of higher institutional authority.

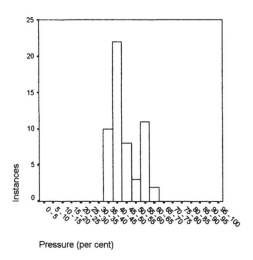

Pressure (per cent)

Figure 4.5 Blocked preference changes

In the case of air strikes in Bosnia, it could thus be observed that during the initial phases of the debate all major international organizations, such as the WEU, the EC Council of Ministers, the North Atlantic Council and the UN Security Council, were involved in the negotiation of an international response to the crisis in the former Yugoslavia, giving a maximum range of actors a veto. Because of disunity within the WEU, the EC and NATO, however, the French government succeeded in transferring the authority over the international response to the conflict in Bosnia to a higher international decision unit, the UN Security Council, by summer 1992.

Although the change in the ultimate decision unit limited the ability of most European governments to determine the nature of the international response to the conflict in the former Yugoslavia, they accepted the transfer in return for decreased responsibility. It enabled most of them to stand aside, while two Security Council members, namely France and Britain, provided the majority of ground troops for the United Nations protection force UNPROFOR. In return, the change of the ultimate decision unit decreased the likelihood of a decision in favour of military action for the permanent UN Security Council members, who were unanimously opposed to air strikes at the time of the transfer.

In most international organizations, the member governments respected the transfer of the authority to the UN Security Council and refrained from taking further decisions on the issue. Although the members of NATO and the WEU were increasingly divided over air strikes, further action on their part was repeatedly made dependent upon prior authorization by the Security Council. Nevertheless, member states used the linkages of international organizations to exert further pressure on the Security Council. Thus, the EC Council of Ministers called upon the international community to consider military action on 23 March 1993. Moreover, a meeting of foreign and defence ministers within the WEU in May paved the way for the implementation of air strikes to protect the safe areas around Sarajevo, Tuzla, Zepa, Bihac and Gorazde.

Table 4.1 Descriptive statistics

Preference changes	Number of instances	Range of pressure	Minimum pressure	Maximum pressure	Average pressure
No change (NC)	178	52%	0%	52%	30%
Unclear or undecided (U)	45	64%	17%	81%	47%
Change (C)	19	80%	20%	100%	43%
Blocked (B)	56	28%	35%	64%	46%

A third measure for the correlation between the degree of pressure and preference changes is the difference in the average degree of the pressures in each behavioural category. Table 4.1 shows that actors who maintained their opposition to air strikes in Bosnia were on average not exposed to pressure above 30 per cent. However, those who changed their policy preferences in favour of military action were on average pressed by 43 per cent of the actors to whom they were linked in the multilevel European foreign policy network. Actors who were unclear or undecided about their position towards air strikes in Bosnia were on average exposed to even higher degrees of pressure at 47 per cent, as well as those actors in which a veto prevented a policy change at 46 per cent.

Important for a positive evaluation of the two hypotheses the findings corroborate that actors who maintained their policy preference were more often subject to low degrees of pressure than those who changed their preferences. Moreover, in accordance with the second hypothesis the average degree of pressure which actors were able to withstand was higher among those who were blocked. Concerning the category of unclear or undecided preferences which is not specified in either hypothesis, the statistical evidence seems to confirm the inductive proposition that this type of behaviour also enables actors to resist higher average degrees of pressure.

The final indicator for the plausibility of the first hypothesis is the timing of the preference changes during the research period. It shows a strong link between increases in the degree of pressure and subsequent changes in the policy preferences of the actors. Notably in the previous analysis, ten out of fifteen actors, namely the media, the members of the US-Congress, the EU Commission, the British electorate, the staff of the US White House, the British Foreign Secretary, the members of the Labour Parliamentary Party, the British Defence Secretary, the British Prime Minister, the British Cabinet, the Chancellor of Exchequer, the WEU and the North Atlantic Council changed their preferences immediately after the pressure on them had increased. Two actors, the American President and the UN Security Council, modified their preferences in the second phase after the pressure on them had increased. Four role actors resisted an initial strengthening of the demand for air strikes, namely the US President, US Secretary of Defence, the EU Council of Ministers and Foreign Office officials. However, the ability of these actors to resist the pressure appeared to be due to the fact that two of these four actors were subject to no further increases until they changed their preferences. Moreover, the degree of pressure on each of them was below the 50 per cent threshold identified above. The time span over which actors were able to withstand the demands forced upon them varied considerably, ranging between two and six phases.

Conclusion

A number of conclusions are suggested by the preceding analysis with regard to the making of European foreign policy within the transatlantic community. Specifically, this chapter has illustrated the important role played by NATO and the United Nations in defining European foreign policies. Moreover, it has demonstrated the close relationship between Europe and the United States which shapes foreign policies on either side of the Atlantic. This chapter thus confirms the contention presented at the beginning of this book that the multilevel European foreign policy network stretches beyond Europe and the European Union.

The density of the relations between these multiple actors is very much reflected by the interactions observed in the preceding case study. Thus, multilevel network theory illustrates that pressure for the permission of air strikes in Bosnia was exerted nationally, transnationally and internationally. At the international level, ministers and officials from Germany, Italy, Turkey, Portugal, the Netherlands, Belgium and Austria in particular lobbied for air strikes. In order to do so, they used their membership in international organizations, such as NATO, the WEU and the EU, as well as transgovernmental relations with their colleagues in the Foreign and Defence Ministries in the transatlantic community. From autumn 1992 additional transnational pressure emerged from the members of the international media and the US Congress, while nationally the British government, for instance, came under pressure first from Liberal and later from Labour MPs in the House of Commons.

.The multiplicity of the relations through which the advocates of air strikes in Bosnia could and did seek to influence the United States and British administrations disconfirms the notion of the government as a gatekeeper between the national and the international arena. Particularly interesting is the observation that one of the transnational coalitions which emerged in opposition to military action in Bosnia involved staff from the Pentagon, the British Ministry of Defence and the military within NATO which publicly challenged the policies of their own political leaders. The action shows that governments can by no means be perceived as unitary actors as the gatekeeper concept implies. Conversely, the case study indicates that politicians and civil servants equally used their relations within the multilevel European foreign policy network to further their own, not necessarily congruent, foreign policy preferences.

In addition to demonstrating the transatlantic scope of the European foreign policy network and the possible coalitions which emerged within it, multilevel network analysis provides new insights into the foreign policy decision-making process. In accord with the first hypothesis this case study reveals how foreign policy and preference changes can be explained by the progressive increase of pressure on particular actors as the result of shifting preferences among those actors to whom they are linked in the multilevel European foreign policy network.

Furthermore, the case study illustrates how vetos and blocking behaviour in collective organizations, such as the WEU, the OSCE, the North Atlantic Council and the UN Security Council, can shape the decision-making process and its outcomes. Two actions in particular were the result of a strategic use of veto behaviour by key actors. The first was the exclusion of the OSCE from involvement in the Yugoslav crisis and perhaps also a greater role in the Post-Cold War European security architecture due to the veto of the former Yugoslav government and Russia. The second was the transfer of the ultimate authority over the international mission in the former Yugoslavia to the UN Security Council because of the respective British and French blocking behaviour in the EU, the WEU and NATO. Thus, although an international coalition of Germany, Italy, Turkey, Portugal, the Netherlands, Belgium and Austria was able to raise the pressure for air strikes within these international organizations to a considerable degrees, military action was prevented in autumn 1992.

The blocking of a UN Security Council decision in favour of air strikes until June 1993 also illustrates how important the sequence of interactions and preference changes is for the definition of foreign policy outcomes. Specifically, the preceding analysis indicates that the pressure of the initial proponents of military intervention was insufficient in bringing about a preference reversal among key actors in the summer of 1992. All major international organizations were blocked from taking action by vetos from Britain, France and the US. Moreover, the transgovernmental pressure on British ministers and officials was too low to overcome persistent majority against military action among British MPs and the MoD. Thus, it was only after US President Bush and his administration changed their position on air strikes following news of Serb concentration camps, that the decision-making process gained new impetus due to US pressure for air strikes.

Finally, although a winning coalition is by definition confined to actors directly linked to a particular decision unit, a multilevel network analysis of the case of air strikes in Bosnia suggests that it was not a coincidence that the decisions of the American and British governments to endorse air strikes each came about at a time when air strikes were supported by a majority of domestic and international actors to which the two administrations were linked. The findings thus indicated that single-level models which focus on domestic, transnational or international factors would have equally been able to explain the decisions of the US and British governments to support air strikes. However, by restricting themselves to one level of analysis they would have missed an important aspect of the decision-making process in the convergence of preferences among national and international actors.

Notes

1. In the following 'Bosnia'.
2. Altogether 16 actors were linked to the 651 MPs in the House of Commons, accounting for $L=16 \times 651=10416$, among these 20 liberal MPs used their links to exert pressure in the Commons as well as on other MPs, raising E to $E=320+631=951$ and P_1 [Par] $=E/L= 951/10416 = 9\%$.
3. Eleven of 56 actors to whom the national and international media had links in the network favoured air strikes, namely the governments of Germany, Italy, Turkey, Portugal, the Netherlands, Belgium and Austria, US Secretary of State James Baker, US National Security Advisor Brent Scowcroft, officials in the US State Department and the members of the Liberal Democrats, accounting for P_1 [Med] $= 11/56 = 20\%$.
4. In the case of Defence Secretary Rifkind, six actors advocated air strikes among his 29 linkages, namely the defence ministers of Germany, Italy, Turkey, Portugal, the Netherlands and Belgium, raising the pressure to P_1 [DS] $= 6/29 = 21\%$.
5. Foreign Office officials were pressed for air strikes by their counterparts in the US State Department and the Foreign Ministries of Germany, Italy, Turkey, Portugal, the Netherlands, Belgium and Austria, accounting for P_1 [Fco] $= 8/30 = 27\%$.
6. Similarly, the British Ministry of Defence staff was pressurized by their defence ministry colleagues in Germany, Italy, Turkey, Portugal, the Netherlands and Belgium, i.e. P_1 [Mod] $= 6/21 = 29\%$.
7. In the Western European Union, the heads of state, foreign ministers, defence ministers, foreign ministry staff and military officials from Germany, Italy, Portugal, the Netherlands and Belgium accounted for P_1 [Weu] $= 25/47 = 53\%$.
8. In NATO's integrated organization, the diplomatic and military staff from Germany, Italy, Turkey, Portugal, the Netherlands and Belgium as well as US State Department officials raised the pressure to 13 out of 29 linkages, i.e. P_1 [Nato-Org.] $= 13/29 = 45\%$.
9. In the European Union Council of Ministers, the same balance applied as to the Western European Union with P_1 [EU-CM] $= 25/63 = 40\%$.
10. In the North Atlantic Council, the advocacy of air strikes by the heads of state, foreign ministers and defence ministers from Germany, Italy, Turkey, Portugal, the Netherlands and Belgium, as well as James Baker accounted for 19 of its 51 linkages in the network, with P_1 [Nato-CM] $= 19/51 = 37\%$.
11. In the CSCE air strikes were supported by 23 out of 65 actors who were linked to the organization, including the heads of state, foreign ministers and diplomatic officials from Germany, Italy, Turkey, Portugal, the Netherlands, Belgium, Austria, the US Secretary of State and State Department staff, with P_1 [CSCE] $= 23/65 = 35\%$.
12. P_1 [UN-SC] $= 1/10 = 10\%$.
13. Due to the changes of opinion among Congress members and journalists the pressure on Bush increased from nine out of 32 actors to 11, i.e. from P_1 [Pre] $= 9/32 = 28\%$ to P_2 [Pre] $= 11/32 = 34\%$.
14. Due to the preference change among Senators, the media and President Bush, ten and seven actors who were linked to Defence Secretary Cheney and Pentagon officials respectively favoured air strikes, raising the pressure on the former to P_3 [US-SD] $= 10/24 = 42\%$ and the latter to P_3 [US-pen] $= 7/19 = 37\%$.
15. P_2 [EU-Co] $= 11/27 = 41\%$.

16. Specifically, nine out of 39 actors to whom Major had regular contacts favoured air strikes, including the heads of state in Germany, Italy, Turkey, Portugal, the Netherlands and Belgium, the international press and the EU Commission, accounting for P_4 [PM] = 9/39 = 23%.

17. James Adams, Louise Branson, Ian Glover-James and Andrew Hogg 'Have they got away with it?', *Sunday Times*, 16 August 1992.

18. The pressure on the North Atlantic Council had increased due to President Bush's preference change by one out of 51 from P_3 [Nato-CM] = 20/51 = 39% to P_4 [Nato-CM] = 21/51 = 41%.

19. P_6 [Fco] = 8/30 = 27%.

20. The preference change among Foreign Office officials increased the pressure on the Foreign Secretary by one from nine to ten out of 36 actors, i.e. from P_6 [FS] = 9/36 = 25% to P_7 [FS] = 10/36 = 28%.

21. Specifically, the support for air strikes from Foreign Secretary Hurd increased the pressure on the members of the two parties from P_7 [con; lab] = 2/12 = 17% to P_8 [con; lab] = 3/12 = 25%.

22. In addition to their counterparts from seven European countries, US President Bush, the international media, now Foreign Secretary Hurd and Foreign Office staff supported air strikes, raising the pressure on Prime Minister Major to P_8 [PM] = 11/39 = 28% and on Defence Secretary Rifkind to P_8 [DS] = 8/29 = 28%.

23. The preference change among Foreign Office staff increased the pressure on their MoD colleagues by one to P_8 [Mod] = 7/21 = 33%.

24. Among the actors linked to the House of Commons four, namely Foreign Secretary Hurd, the media and the members of the Liberal and Labour Parties increased the pressure to P_{10} [Par] = 6096/10416 = 59%.

25. The change of public opinion increased the pressure on Rifkind from nine to ten out of 29 actors to which the Defence Secretary was linked., i.e. from P_{11} [DS] = 9/29 = 31% to P_{12} [DS] = 10/29 = 34%.

26. The pressure on Prime Minister Major increased with the electorate by one out of 39 actors from P_{11} [PM] = 12/39 = 31% to P_{12} [PM] = 13/39 = 33%.

27. The pressure on the Cabinet increased from three to four out of 14 actors, i.e. from P_{11} [Cab] = 3/14 = 21% to P_{12} [Cab] = 4/14 = 29%.

28. The preference change of the Defence Secretary raised the pressure on the Prime Minister by one to P_{13} [PM] = 14/39 = 36%.

29. P_{14} [Cab] = 6/14 = 43%.

30. Specifically, the heads of state, foreign ministers, defence ministers, foreign ministry staff and military officials from the UK, France, Germany, Italy, Portugal, the Netherlands and Belgium exerted pressure on the Western European Union, accounting for P_{15} [Weu] = 35/47 = 74%.

31. In the UN Security Council the US, Britain and France now favoured air strikes, Russia did not support military action, but decided not to veto the resolution in the light of the failure of the Vance-Owen peace plan, thus raising the pressures to P_{16} [UN-SC] = = 60%.

32. Within NATO air strikes were now supported by the US, Britain, France, Germany, Italy, Turkey, Portugal, the Netherlands and Belgium with additional pressure from the UN Security Council and P_{17} [Nato-CM] = 28/51 = 55%.

33. The measure excludes blocking behaviour because it is explained by the second hypothesis.

5   United Kingdom:
The Tactical Air-to-Surface Missile

Introduction

While the preceding chapters have illustrated that European foreign policies are increasingly influenced and even implemented by international organizations like the European Union and NATO, national governments remain key actors in making of authoritative foreign policy decisions. In particular in the area of national security, governments have sought to safeguard their sovereignty.

This chapter examines to what degree national European foreign policies in the sensitive area of national defence are, nevertheless, determined by the pressures and interactions which characterize policy making within the multilevel European foreign policy network. As an example, it analyses the cancellation of the British tactical air-to-surface missile (TASM) project in 1993. In particular, this case study demonstrates how the pressure from Britain's European partners not only led to a re-evaluation of tactical nuclear missiles within NATO, but also contributed to this change in British defence policy.

The debate over the British development of a tactical air-to-surface missile arose with the end of the Cold War. As democratic reforms spread over Eastern Europe, existing nuclear defences aimed at these countries appeared obsolete, if not counterproductive (Boutwell, 1990: 218). In particular, the Lance missile system with its 300-mile radius did not conform to the changed international environment. In May 1990, a review of NATO's nuclear strategy was set up to consider tactical alternatives as the basis for a new nuclear defence posture in Europe. During the review it emerged that the majority of NATO member states opposed the replacement of the short-range Lance by TASMs as had been proposed by the governments of the United States and Britain. Yet, in spite of the international opposition, the British government proceeded with the examination of TASMs as part of its independent nuclear deterrent and as a contribution to NATO's nuclear defences. Over the following three years, the British government specifically considered two alternatives: a collaborative development of the missile with the French Aerospatiale company and two off-the-shelf platforms from the American companies Boeing and Martin Marietta. However, due to international pressure the support for a new nuclear missile within the British administration diminished steadily. By August 1992 the TASM had effectively lost all backing within the MoD and the Cabinet. Nevertheless, only after a year of conspicuous silence, Defence Secretary Malcolm Rifkind announced the cancellation of the TASM programme on 18 October 1993.

Due to two conditions, the case appears particularly appropriate for the testing of multilevel network theory. First, as in the cases of dual-use export controls in Europe and air strikes in Bosnia, exogenous factors cannot sufficiently explain the British foreign policy change. In particular, the end of the Cold War was considered by no means as undermining the requirement for tactical nuclear weapons. In fact, some actors argued that the breakup of the Warsaw Pact and later the Soviet Union enhanced the need for sub-strategic nuclear weapons because they were more suitable for the new threats in form of small, but volatile states in Eastern Europe, the Mediterranean and the Middle East.

In addition, the application of multilevel network theory to this case is intriguing because the argument put forward by the British government for its cancellation of the TASM project, namely that the decision had been due to budgetary pressures, does not fully explain the change in British nuclear policy (Rai, 1993). The timing of the cancellation in October 1993 remains especially incomprehensible. Pressures to reduce defence spending had already emerged in the late 1980s and were reinforced by the end of the Cold War in 1990 (Carver, 1992; Croft and Dunn, 1990). Nevertheless, the British government resisted them at the time by claiming that the changed nature of the international system required a new tactical nuclear weapon. Moreover, the government made the unpopular decision to reduce its standing forces which involved substantial job losses in sensitive regions across Britain in order to free budgetary resources for a continuation of the TASM project. In fact, as late as September 1992, the government invested another 4.8 million dollars in pre-project studies for a TASM by the US companies Boeing and Martin Marietta and France's Aerospatiale. Furthermore, it can be contended that, if budgetary reasons had been the primary concern of the government, the TASMs would have been preferable to a fourth Trident, which was chosen as sub-strategic alternative to the TASM in 1993. Not only was the TASM cheaper, it was, according to military experts, also more suitable for British defence (Paterson, 1997: 113). In fact, the British government itself had argued along these lines in November 1992 and again in 1993.

The Abandonment of the Tactical Air-to-Surface Missile

The reassessment of TASMs arose with the end of the Cold War. Although the weapon system had always been controversial within the Atlantic Alliance, national and international doubts over the necessity of the missile increased considerably after the dissolution of the Warsaw Pact. Most European governments agreed that the changed nature of the international system reduced the need for large nuclear arsenals. Although few governments questioned the continued relevance of nuclear defences in principle, the development of new nuclear weapons systems appeared to contradict the dismantling of old stocks by NATO and the former Warsaw Pact countries. Moreover, the increasing development cost of new weapons weighed heavily on national defence budgets which were drastically cut back in the early 1990s to meet popular demands for a peace dividend.

Still at the pre-development stage, the TASM project was particularly vulnerable to the calls for cutbacks. The investigation into a TASM had been begun by the United States in 1987 without the consultation of its European allies. At the time, the project had already caused some controversy in NATO because the governments in Belgium and Germany challenged the requirement for new nuclear weapons in Europe. Only the British administration, which sought a replacement for its WE-177 free-fall nuclear bombs, was keen to purchase the missile. However, the British government also investigated a collaboration with France in the design of a TASM (Rai, 1993: 15). In November 1989, the British Ministry of Defence awarded one million pounds to the French Aerospatiale company for a pre-feasibility study. Apart from political reasons, problems with the development of nuclear warheads at the Aldermaston Atomic Weapons Establishment increased the attractiveness of the Franco-British option. While the US military was prohibited from sharing certain information about nuclear development under the US Atomic Energy Act of 1959, a collaboration with France was not legally restricted. However, a Franco-British project could endanger British nuclear testing in the US sites in Nevada. Moreover, Royal Air Force (RAF) staff preferred the US American options which included Boeing's SRAM-T and Martin Marietta's Tactical Air-to-Surface Missile. In May 1990, both options were still under investigation. As the criticism of the project increased, the focus of the decision-making process shifted from the selection of the missile to the question of whether TASMs were required at all. Moreover, in the latter stages of the debate, the issue was redefined as a direct competition between TASMs and a tactical missile to be carried by Britain's Trident submarines. This case study examines how the interactions among the members of the multilevel European foreign policy network shaped the debate for and against the TASM.

Opposition to TASM in NATO Strategic Review

The international debate over TASMs began in early 1990, when the US President George Bush proposed the abolition of plans to replace Lance as a starting point for US-Soviet negotiations to remove all short-range nuclear weapons from Europe. In addition, Bush initiated a review of NATO's nuclear strategy in a summit proposed for June or July. In spite of these measures, American and British political leaders agreed that the US should maintain its nuclear presence in Europe (Carver, 1992:165). Both governments planned to deploy the air-launched TASM as replacement for the land-based Lance (Wheeler, 1990: 43). In the US, the TASM was praised as a weapon which would be suitable for the changed international security environment after the end of the Cold War. Nevertheless, other actors within the multilevel European foreign policy network soon voiced their opposition to the missile. Specifically, the German administration objected strongly to the British and American plans for basing the TASM on the European continent. Following the dismantling of the Lance the German Foreign Minister Hans-Dietrich Genscher had hoped for the denuclearisation of Central Europe. The opposition of the German administration to the missile was shared by several of its European neighbours, such as the Netherlands, Belgium and, to some degree, Italy. In addition, the development of new nuclear weapons was criticized by representatives of the British Labour Party and the Campaign for Nuclear Disarmament (CND).

At the beginning of May 1990, however, the opposition to the TASM had yet to organize itself. The linkages within the network through which the four governments and the two domestic opponents of the missile could exert pressure for the abolition of the missile programme were restricted. The strongest relations existed between the foreign and defence ministers of Germany, Belgium, the Netherlands, Italy and Britain, and the officials from their respective ministries. However, combined, the politicians and civil servants of four countries did not amount to more than 28 per cent of the actors to whom the British Secretary of Defence, who had the primary authority on the issue, was linked in the network.[1] Moreover, the four administrations did not have any contacts with the domestic opposition to the TASM in Britain among the Liberal Democrats, the Labour Party and the CND with whom they could have formed a transnational coalition.

The best prospects of increasing the pressure on the governments of Britain and the US existed within NATO which was the only international organization in Europe with a significant role in nuclear decision-making. Since the TASM had been made a requirement under NATO's integrated nuclear strategy in the 1980s, a common review of the Alliances strategic defence posture could lead to a reevaluation of the missile. Indeed, the defence ministers from the four governments used the next meeting of NATO's Nuclear Planning Group (NPG) in Kananaskis, Canada, on 9-10 May 1990, to express their criticism of the TASM. Predictably the four ministers clashed with the British Defence Secretary Tom King who had hoped to gain the member states' approval for the deployment of the TASM on the continent. Questioned by journalists, NATO General Secretary Manfred Wörner admitted that the policy faced not only the opposition from many European governments, but also the Soviet Union which demanded a nuclear-free Europe as condition for its consent to German reunification.

Further discussions in the NPG revealed the scope of the opposition to the missile in NATO. Not only the representations of Germany, the Netherlands, Belgium and Italy criticized the proposal to station TASMs in Europe, but also Denmark, Iceland and Norway. The defence secretaries of the US and Canada spoke out in favour of the TASMs. However, their countries were not suited for deployment if the missile was to be used in conflicts in Eastern Europe or the Middle East. This left only the British and, perhaps, the Turkish government prepared to base the missile in their country. The governments of France and Spain did not object to the TASMs, but because they were not integrated into the military structure of NATO, they had no voice in the NPG. With seven NPG members explicitly opposed to the stationing of TASMs in Europe, 41 per cent of the actors linked to and represented within NATO exerted pressure against the missile.[2]

NATO's TASM Cancellation Blocked

However, in spite of the strong opposition, the international coalition among the defence ministers from the seven countries failed to achieve the unequivocal cancellation of the programme in the North Atlantic Council. The factual veto position of all NATO member states enabled the British and American representatives to block a change of NATO policy at this point. Crucial for the success of this strategy was the fact that the requirement for a tactical nuclear missile as part of NATO's nuclear defence strategy had already been approved in the 1980s. The abandonment of TASMs was, therefore, a change from the *status quo* which Britain and the US could veto. The blocking position of the two governments in the Alliance effectively increased the pressure required for a policy change in the Alliance. Furthermore, the opposition to the missile was internally divided. Although not in favour of the missile, most NATO members had not yet developed a clear policy preference with regard to the future of NATO's nuclear defence strategy.

In particular, the German administration would have preferred to avoid the topic of nuclear weapons altogether because it feared that its discussion would endanger the ongoing 'two-plus-four' negotiations with the Soviet Union over German reunification. Moreover, German politicians were aware that a public debate over the stationing of nuclear missiles in Germany would almost certainly lead to public protests as occurred after the deployment of Cruise and Pershing missiles in the 1980s (Williams, 1990: 20). Further indications of unclear policy preferences were differences in the governmental policy as expressed by the German Foreign Office and the Defence Ministry. While Foreign Minister Genscher and his staff objected to the replacement of nuclear missiles in Germany, Defence Minister Gerhard Stoltenberg avoided taking a critical stance on the issue of the TASM. Chancellor Kohl's resistance to the deployment of new nuclear weapons in Germany appeared to be due more to its effect on the negotiations over German reunification than to fundamental objections to nuclear weapons. Kohl certainly disagreed with Foreign Minister Genscher's advocacy of a denuclearized Europe. To avoid a controversy with his national and international partners, Chancellor Kohl argued that it was too early to decide about the issue (Boutwell, 1990: 229). Even the American intentions were unclear. Although President Bush had suggested a review of NATO nuclear strategy for the summer, Defence Secretary Richard Cheney stated at the meeting of the NPG that NATO's flexible response doctrine would not be put into question.

NATO Decision Postponed

In order to allow for a policy review both internally and among its member states, the NPG decided to postpone the issue. In the NPG's Final Communiqué of 10 May 1990, the conflicting views among its members over the future of nuclear weapons in Europe were played down. The TASM programme was not mentioned. The communiqué stated, however, that due to a reduction of short-range nuclear missiles, sub-strategic systems would 'assume relatively greater importance'. If this suggested that the British and American delegates had prevailed in the discussion, their victory was built on shaky grounds because the NPG also agreed to task General John Galvin, NATO's Supreme Allied Commander in Europe (SACEUR), with a comprehensive review of conventional and nuclear weapons requirements after the end of the Cold War. The review would enable the opponents of the missile to express their concerns and co-ordinate their pressure along their national, transnational and international relations within the network over the coming months. In particular, foreign and defence ministers and officials from the seven countries were able to use the transgovernmental linkages with their British and American counterparts to press for a review of the TASM programme.

Thus, the transgovernmental pressure on the American administration was considerable. The degree of pressure was, at 37 per cent, highest on the relevant officials in the Pentagon who had close relations with their partners bilaterally and through NATO, but few other linkages within the network.[3] The American Secretary of Defence Cheney was also subject to considerable pressure at 33 per cent,[4] while President Bush was less affected as the seven heads of state only accounted for 25 per cent of his linkages.[5] In Britain, the pressure on the administration was equally high at the level of the Ministry of Defence staff where the counterparts from the eight countries accounted for 33 per cent of their contacts in the network. However, the pressure from the eight administrations was significantly lower on Foreign Secretary Douglas Hurd at 22 per cent and Prime Minister Margaret Thatcher at 21 per cent.[6] Crucially, the British Cabinet was only indirectly linked to the governments through the departmental ministers and, therefore, insulated from international pressure as long as British ministers refused to abandon the TASM.

Britain and US Continue TASM Development

The lack of direct pressure on the Cabinet helps to explain why, in spite of the international protests, the British government continued to support the TASM programme during autumn 1990. In the House of Commons, Minister of State for Defence Archie Hamilton emphasized the government's determination to proceed with the deployment of TASMs regardless of the opposition from its European partners. Speaking to the Commons Defence Committee, three former military service chiefs and a former civil servant from the Ministry of Defence pointed out that in order to maintain a viable defence, it was essential for Britain to retain its nuclear deterrent. While they endorsed the abolition of short-range nuclear weapons by the superpowers, they recommended the deployment of medium-range missiles such as TASMs in their place. The American administration, too, decided to proceed with the development of the missile for the time being. At the end of May it awarded a 181 million-dollar follow-on contract to Boeing Aerospace and Electronics for the full development of the SRAM-T.

In May 1990, the main question for the British government was not whether to proceed with the development of the TASM, but with whom to collaborate. MoD officials were still evaluating its two options: the purchase of American missiles or to cooperate with France. Prime Minister Thatcher increasingly appeared to support the French option. As a result of bilateral talks with President Mitterrand on 7 May, Thatcher agreed to enhance the cooperation between the British and French armed forces. The French government was little affected by the opposition to the missile within NATO and, thus, would be a reliable partner in the development of a TASM. While the US and Britain were increasingly under pressure in NATO, the French administration was unconcerned about the debate in the Atlantic Alliance as France had left NATO's integrated military structure and maintained its nuclear independence. Moreover, the French government agreed with its counterparts in the US and Britain over the need to retain nuclear forces in Europe. In the view of the French administration, the end of the Cold War did not question the necessity of an independent nuclear deterrent for France. On the contrary, as the potential nuclear threat moved from the Warsaw Pact to the Soviet Union, the Middle East and Third World countries, the French military had identified the TASM as a key element of its new defence posture (Larkin, 1996: 26).

Although the British and American representations had temporarily prevailed in the North Atlantic Council and resisted international pressures for the abolition of the TASM, both governments recognized that it was crucial to gain the active support of their European partners if they planned to deploy the TASM in Europe. In order to do so, the British and American military began in turn to exert pressure on their colleagues for a positive reevaluation of the TASM. Soon after the NPG meeting in May, information was leaked from officials in Bonn and Washington about a Whitehall plan to overcome German objections to the TASM. According to a proposal discussed between senior NATO and Pentagon officials, a joint British, American and German air force based in Britain could integrate the German Bundeswehr into NATO's nuclear defences, yet avoid the stationing of the missile on German soil. The plan was a step back from the original British intentions to station the TASM on the continent. Nevertheless, when Foreign Secretary Hurd raised the question during a meeting in Bonn, the German Foreign Minister Genscher immediately rejected the plan. Further discussions were envisaged for a gathering of NATO defence ministers in the Defence Planning Committee later in the week, but the plan was quietly dropped.

Soviet Union Joins TASM Opposition

In June 1990, the leadership of the Soviet Union used the discord within NATO to add to the transgovernmental pressure on the British and American administrations for the abolition of the TASM programme to achieve its own ends. Soviet representatives suggested the inclusion of British and French tactical nuclear arsenals into the disarmament negotiations with the US. Soviet officials argued that TASMs and sea-launched cruise missiles effectively undermined the Intermediate-Range Nuclear Forces (INF) Treaty which had been signed in 1987. The treaty applied only to ground-launched nuclear weapons. The claim received unexpected support from US Admiral Eugene

Carroll who admitted that the modernization of NATO's nuclear forces was 'in breach of the spirit, if not the letter, of the INF Treaty'.[7] In Britain, representatives of the Labour Party and the CND made a similar argument. Nevertheless, politicians in the US and even Germany, which was opposed to the TASM, rejected the suggestion because of its possibly far-reaching consequences.

At the next meeting of NATO foreign ministers in Turnberry, Scotland, on 7-8 June 1990, the distribution of preferences among the member states was unchanged. The British Prime Minister Margaret Thatcher reiterated her intentions to station TASMs in Britain and other European countries in order to maintain NATO's nuclear defences. She was again challenged by the German Foreign Minister Hans-Dietrich Genscher who was informally quoted as having said that Germany would refuse outright the deployment of TASMs if asked. An open confrontation between them was narrowly avoided when Genscher denied the statement. However, German representatives at NATO repeatedly made clear that their government was not in favour of basing TASMs in Germany. Again NATO ministers agreed to postpone a decision regarding the TASM - this time until 1992. Although the degree of pressure on the Alliance had not changed, its consistently high level of 41 per cent began to wear down the ability of the American and British foreign ministers to veto the abolition of the TASM requirement. In the Turnberry communiqué, the Alliance position was certainly formulated more carefully than it had been in May. In particular, NATO member states now expressed their willingness to consider and initiate adjustments in their number of conventional and nuclear forces.

Soon after the NATO meeting, the Soviet leadership proposed to abolish all short-range nuclear arsenals and to cancel the French and American development of TASMs in negotiations which were to start in September. However, the French government rejected the suggestion. It fundamentally refused to negotiate its nuclear defences under a comprehensive multilateral framework together with those of the US and Britain. The explicit position of the French President Mitterrand was that his administration would not join the international negotiations before the nuclear arsenals of the two superpowers had reached levels comparative to that of France (Larkin, 1996: 139-141). In the meantime the French government would continue with the development of its new tactical nuclear missile. In fact, French and US scientists had carried out nuclear tests for the design of TASMs only days before the Soviet announcement. The North Atlantic Council also rejected the Soviet proposals, but agreed on negotiations over short-range nuclear forces with the Soviet Union to begin in 1991. Moreover, the persisting international pressure from the representatives of many NATO member states and the Soviet Union for the abolition of the TASM showed first signs of weakening the American government's resolve to update its tactical nuclear weapons.

NATO Nuclear Weapons 'Last Resort'

The issue reemerged during the review of NATO's nuclear doctrine at the summit in London on 5-6 July. In a letter to Alliance leaders, the American President George Bush suggested making NATO's tactical weapons means of last resort. In addition, Bush pressed for a common position on the reduction of short-range nuclear forces. While US Secretary of State James Baker agreed with his British colleagues that the Alliance should reserve its right to respond to a conventional attack with nuclear missiles, his government seemed increasingly divided. The split ran between the hawkish Defence Secretary Richard Cheney and Vice-President Dan Quayle, and the cautious President Bush. To avoid the public debacle of the previous meetings, NATO heads of state agreed not to discuss the contentious TASMs at the summit. However, this informal agreement did not prevent the British Prime Minister Margaret Thatcher from raising the subject of the missile in her speech. Thatcher pointed out that the Soviet Union continued to build 100 TASMs per week implying that NATO should respond in kind. Thatcher also opposed President Bush's proposal to make tactical nuclear weapons means of last resort. Conversely, the German Chancellor Helmut Kohl supported the American initiative. Moreover, Chancellor Kohl himself privately dismissed the deployment of TASMs in Germany.

The eventual endorsement of President Bush's proposal by NATO leaders was perceived by the media as a surprise considering the unabated opposition of the British Prime Minister to a reduction in NATO's nuclear defence posture. However, the degree of international pressure on the members of the North Atlantic Council can account for the erosion of NATO's commitment to tactical nuclear weapons. Although the veto position of Britain and the US enabled Prime Minister Thatcher and President Bush to block the outright cancellation of the TASM requirement in the Council, the persistent pressure from 41 per cent of NATO's members was forcing progressive shifts in its nuclear policy. In their Final Communiqué of 6 July 1990, NATO heads of state declared that nuclear forces would be made 'truly weapons of last resort'. Although British officials tried to play down the associated change in NATO strategy, the new policy implied that its nuclear defence would now rest on long-range strategic weapons. It put the future role of the middle-range TASM implicitly into question.

Opposition to TASM Increases in Britain

Following the defeat of the British administration on the question of making nuclear weapons means of last resort in NATO's strategy, the domestic opponents of the TASM in Britain, namely members of the Labour Party and the CND, reasserted their criticism. In the absence of direct influence on the government they focussed their pressure on the national and international media. Spokespersons for the Labour Party used the public attention created by the NATO summit to expressly welcome NATO's decision. The new NATO strategy matched the preference of Labour members for a no first-use policy of nuclear weapons. Moreover, party leader Neil Kinnock expressed his scepticism about the development of the TASM in the American press during a visit to Washington. However, Labour leaders indicated that they would honour British commitments if NATO decided to base the missiles in Europe. In Labour's 1990 party programme, 'Looking to the Future', the Labour Party appeared similarly divided. While the programme advocated the destruction of all land-based short-range nuclear weapons, it did not discuss other nuclear weapons such as the TASM. Labour leaders themselves were under pressure from representatives of the CND among their members. CND representatives intended to use the issue of the TASM as a lever for a broader critique of NATO policy after the end of the Cold War. They planned a campaign against the deployment of the TASM for the beginning of 1991. Moreover, CND members could claim to have broad support from the British public. According to a survey, which the CND had commissioned from Gallup, 60 per cent of the population were opposed to the TASM. Members of the Liberal Democrats also criticized the government for its plan to go ahead with the purchase of a TASM. In a political paper for its party conference, titled 'Reshaping Europe', Liberal Democrat leaders expressed their opposition to the replacement of the WE-177 free-fall nuclear bomb by a TASM.

However, the ability of opposition MPs and members of the CND to exert pressure within the European foreign policy network was very limited. In particular, they lacked direct leverage over cabinet ministers or officials. In order to influence the administration, they had to use indirect linkages through the House of Commons, the members of the Parliamentary Defence Committee and the media. In Parliament, a preference change was ruled out due to the blocking position of the Conservative parliamentary party. Thus, although the opposition members collectively raised the pressure on the Commons to 46 per cent, this was not sufficient to overcome the dominance of the Conservative parliamentary majority.[8] A change of the Parliament's support for the TASM required first a preference change among the Conservative MPs. However, this was highly unlikely. Not only were Conservative MPs the main supporters of the TASM, they were also insulated from the international and national pressure against the project because the opponents of the TASM lacked direct relations with the members of the Conservative parliamentary party.

In the Commons Defence Committee, the Conservative majority was equally dominant. However, the balance of pressures in the committee was at 6 per cent slightly more favourable for the opposition parties as a result of the small number of members in the committee and their restricted linkages within the network.[9] In their Tenth Report (Defence Committee, 11 July 1990), the members of the Defence Committee, thus, recognized that the deployment of the missiles in Germany or other NATO countries was in doubt. In the light of the opposition within the Alliance, the report concluded that there was 'no evident urgency to deploy a nuclear-armed TASM: rather the opposite' (p.xxxv). Moreover, the Defence Committee members suggested that a decision over its stationing 'should be taken by NATO as a whole in the light of arms control negotiations' (p.xxxv). Nevertheless, the committee members demanded that a credible nuclear deterrent, including a spectrum of sub-strategic weapons, should be retained. In another paragraph, the members of the Defence Committee supported the technical development of a TASM, perhaps in closer cooperation with France.

The British electorate had theoretically the widest influence within the European foreign policy network because its voting power over both government and MPs. However, on the question of the TASM public opinion was unclear. The government believed that the electorate supported a strong policy on national defence. This view was based on opinion polls which suggested that the public remained in favour of Britain's independent nuclear deterrent (Simpson, 1991: 65). Conversely, a CND poll suggested that voters were increasingly critical of tactical nuclear missiles. Given the different phrasing of the questions and the highly technical nature of the issue, the lack of a clear opinion was little surprising. In the event, however, the lack of a clear preference meant that public opinion did not exert pressure on the government in either direction.

Franco-British TASM Collaboration Likely

In October 1990, the British plans for the missile received another setback when the US Congress announced that it would cut the budget for the development of the American TASM back from 118 to 35 million dollars. In a review of American security policy after the end of the Cold War, congressional leaders had concluded that the TASM was not a priority. US Air Force officials were also increasingly doubtful of the project because of repeated problems with the development of the missile. Although the withdrawal of US government funds limited the choices for British defence planners, it did not generally put the MoD requirement for a TASM into question. Instead, the British government began to redirect its attention to Franco-British collaboration in the construction of a TASM. Following the congressional decision, the British Defence Secretary Tom King met his French colleague Jean-Pierre Chevenement to further explore the possibility of jointly developing a TASM. The French Prime Minister Michel Rocard favoured the collaboration as a step towards an independent European security identity. In addition, Franco-British cooperation would split the costs for the research and development of the weapon. Closer European collaboration in defence procurement was also widely supported among British MPs and the members of the Commons Defence Committee. According to David Owen, MP, it would be 'a historic decision...profoundly important for the development of Europe'.[10] MoD staff, however, were not ready to decide on the question. They were waiting for the results of a feasibility study of a Franco-British missile which were expected by the end of the year.

In the meantime, the international opposition from within Germany, the Netherlands, Belgium, Italy, Denmark, Iceland, Norway and the Soviet Union to the development of tactical nuclear missiles by Britain and the US persisted. After the signing of the Conventional Armed Forces in Europe Treaty in late November, the Soviet Union urged again for negotiations on short-range nuclear weapons, which had been agreed with NATO in the summer. The leadership in Moscow indicated that it would seek the abolition of the TASM project by NATO. The call for nuclear reductions was widely shared among the European public. In particular, the German government was coming increasingly under pressure from the electorate to reject the stationing of TASMs and to declare Germany a nuclear-free zone.

British Cabinet Remains Committed to TASM

However, the opponents of the TASM had so far not been able to enlarge their coalition by pressurizing or persuading any other actors within the European foreign policy network to abandon the project. In particular, the British Cabinet remained free from direct pressure from British ministers and was able to maintain its commitment to the TASM during winter 1990. The resignation of Prime Minister Thatcher, who had been a strong supporter of the missile within the government, did not bring about a change in policy preferences. In December 1990, cabinet ministers reconfirmed their determination to go ahead with the deployment of the TASM. RAF staff were particularly eager to purchase the weapon in order to preserve their role in the nuclear defence of Britain after the phasing out of the WE-177 bomb. Although officials in the

MoD were still considering both the American and French options, Franco-British collaboration received new support due to the interest of Prime Minister John Major in strengthening European relations. The Minister of State for Defence Procurement Alan Clark and his Parliamentary Under-Secretary Kenneth Carlisle were also believed to favour collaboration with France. However, by February 1991 MoD officials who were subject to the highest degree of pressure at 33 per cent because of their linkages with their colleagues in NATO showed first signs of a preference change. After an internal dispute over the selection and the cost of the weapon among the three services, the officials postponed the decision regarding the purchase of a missile platform for another year. It had become increasingly difficult for MoD staff to ignore the effects of the international opposition to the TASM. Crucially, the refusal of Britain's NATO partners to station the TASM on the continent reduced the practical value of the missile for British forward defence against rogue states in the Middle East. In addition, the development cost of the missile was rising because of the congressional cutbacks in the American involvement in the programme. When Navy officials suggested the conversion of Trident missiles for sub-strategic use as a less costly and independent alternative to the TASM, the programme was for the first time openly put into doubt within the British administration.

US Congress to Cut Missile Budget

While the British government maintained its support for the missile in public, the American debate over the future of the TASM project intensified during spring 1991. Driven by countervailing demands from its partners in NATO and American Senators on one hand, and the Secretary of Defence and the American armaments industry on the other, the policy of the US administration began to waver. In February 1991, a US government spokesman announced that the administration planned to withdraw its F-111 aircraft, which would have carried the TASM, from Britain. However, he hastened to add that some of the bombers would be replaced with nuclear-capable F-15E planes. The decision implied that the administration still intended to deploy TASMs in Europe in spite of wide ranging cuts of US forces there. Ironically, the cuts in the defence budget enhanced the commitment of the US government to the TASM deployment which was perceived as a cheap alternative to American soldiers in Europe. The argument was outlined in a report by the Johns Hopkins Foreign Policy Institute in March. In the face of US troop reductions in Germany, the report called for the deployment of TASMs at a limited number of positions in Germany and other countries. The report was supported by leading Senators, such as Democrat Sam Nunn, the Republican chairman of the Armed Services Committee William Cohen, the former Democratic Defence Secretary Harold Brown and the Republican William Simon, former US government officials, military heads and defence experts. However, the widespread publicity of the report could not conceal that the advocates of the TASM were increasingly on the defensive. Most Senators were eager to abolish the missile project.

MoD Divided Over Missile

In Britain, the argument over the choice between the TASM and a sub-strategic Trident increasingly split opinions within the MoD. RAF officers naturally argued the case for preserving a tactical role for the Air Force after the phasing out of the WE-177 in addition to the strategic Trident submarines. In a speech in April, the Chief of Air Staff, Air Chief Marshal Sir Peter Harding affirmed: 'Ministers in this nuclear proliferation world are going to need wider options'.[11] However, during the review of British defence policy before the publication of the annual Defence White Paper on 9 July 1991, RAF staff were more and more on the defensive. In addition to the international opposition to the development and stationing of the missile, budgetary pressures from Treasury officials led to tension between the three MoD services. Nevertheless, cabinet ministers and Conservative MPs remained in favour of the missile, so the cancellation of the TASM was out of the question. Just before the publication of the White Paper, officials from the RAF and the Navy reached an agreement to keep the TASM and leave the strategic Trident as a last resort. The 1991 White Paper 'Britain's Defence for the 90s' (Statement on the Defence Estimates, 1990) maintained that the RAF would 'continue to make our major contribution to the provision of sub-strategic nuclear forces in support of NATO and to provide a national independent sub-strategic deterrent' (p.40). For this purpose, the government was 'studying US and French options to replace [the free-fall nuclear bomb] around the end of the century with a tactical air-to-surface missile to deliver a British warhead' (p.40).

To meet the budgetary demands, MoD ministers and officials decided to cut its civilian staff and the British troops in Germany. The decision was deeply unpopular with MoD troops and service suppliers. However, it allowed the MoD to keep its options regarding the future of the TASM open. The first stage of the debate, thus, ended without a major policy review by any of the members of the multilevel European foreign policy network. However, the international opposition to the TASM had shown some impact on NATO where the British government had to accept a series of decisions which put the future role of the TASM for Europe's nuclear defences increasingly into question. The pressure from many of the continental European governments had also weakened the rationale of the missile for the US government. In the following months this would lead to the first preference changes which resulted in President Bush's cancellation of the American TASM in autumn 1991. The North Atlantic Council and NATO's integrated staff followed suit within weeks.

US Suspends TASM Funds

In August 1991, it became public that the members of the US Congress, who had been critical of the TASM since October 1990, planned to end the government's contributions for Boeing's SRAM-T. The missile programme had already been rejected in the House of Representatives when US Senators temporarily blocked further funds for the development of the SRAM-T. The impending blockage of the administration's contribution to the TASM represented the final straw for Pentagon officials. For more than a year the Pentagon staff had been subject to the highest degree of transgovernmental pressures within the American administration at 37 per cent. Now the military withdrew its support for the missile programme.

The consequences of preference change in the Pentagon for the decision-making process were extensive. Not only did Pentagon staff have the main authority over the decision whether to proceed with the development of an American TASM, the department was also well connected within the multilevel network between the US, Britain and the other NATO members. By using their national and international linkages, Pentagon officials were able to exert pressure on other actors to abandon the programme. Specifically, their support for the cancellation of the missile increased the pressure on the head of their department, Defence Secretary Dick Cheney, from 38 to 42 per cent as Pentagon officials contended that the missile had lost its rationale.[12] Moreover, since Pentagon officials had direct linkages to the President through the Chiefs of Staff, the number of actors in the network who pressed President Bush to abandon the TASM increased from 28 to 31 per cent.[13] The President soon responded to the changed balance of preferences. In September 1991, President Bush announced that the US government would cancel its support for the development of Boeing's SRAM-T as part of his unilateral reductions of nuclear weapons. The policy change of the President pre-empted a congressional vote on the issue. The question whether another missile would be converted to a TASM remained open. Pentagon spokesman Pete Williams commented: 'We have made a decision on that specific system. But we have not made a decision on whether we want to follow along with a TASM-like programme. That is something we want to pursue with our NATO allies'.[14]

NATO Abandons TASM

When the US abandoned their missile programme the balance of pressures in NATO turned against the TASM. Following the preference changes within the Pentagon and by President Bush, 55 per cent of NATO's military links[15] and 47 per cent of the actors directly connected to or represented in the North Atlantic Council[16] were now opposed to the development and stationing of TASMs in Europe. The intergovernmental European coalition among politicians and civil servants against the TASMs used the increased pressure to propose far ranging arms reductions which included the Alliance's stockpile of tactical nuclear bombs. The scope of the reductions was scheduled for discussion at the meeting of the North Atlantic Council in Taormina, Sicily, in mid-October. A formal agreement was expected for the NATO summit in Rome in November. Discussed were two proposals: the removal of all land-based short-range nuclear weapons and cuts of between 1,300 and 1,400 American free-fall nuclear bombs in Europe as well as the British WE-177. After the American government had practically backed out of its TASM programme, representatives within NATO quickly reached a tacit agreement about the cancellation of NATO's requirement for TASMs. Even before the formal meeting of the Council of Ministers in Taormina, a senior NATO official confirmed that the Alliance had decided to suspend its development of TASMs - at least until 1995. The opposition among NATO's members to the TASM programme was simply overwhelming and without the support of the US administration, the British government was unable to veto a policy change any longer. Several NATO member governments welcomed the end of the debate which they believed to be final. The Belgian Defence Minister Guy Come declared: 'The abandonment of TASM is a fundamental change'.[17]

In spite of the agreement, however, the debate within NATO over the future of the TASM continued until the end of November 1991. Central to the discussion was President Bush's wish to maintain an effective nuclear defence in Europe in spite of arms reductions. As replacement for the cancelled SRAM-T programme, the US government considered another TASM: the AM-127 supersonic low altitude target missile. TASMs were still viewed by both US and British ministers as the best option to fill the gap created by the disarmament negotiations. Conversely, the European governments which opposed the TASM demanded the inclusion of the missile in the negotiations. The Soviet President Gorbachev repeatedly expressed his interest in an agreement to remove all tactical air weapons from service. The Soviet proposal was supported by the members of the two main opposition parties in Britain. Labour representatives publicly called upon the British government to use its international role in order to advance the armaments negotiations. Members of the Liberal Democrats demanded in the press that the British government should take the opportunity created by Gorbachev's defence cuts to reduce its military spending by abandoning its TASM programme. The British government had to recognize the growing likelihood that the 'temporary' suspension of the American and NATO requirement of a TASM would be final. Nevertheless, before the NATO Council meeting in Sicily the British Defence Secretary Tom King assured that, whatever the decision in NATO, Britain would go ahead with the deployment of its TASM. Technically, Britain could still pursue both options, i.e. the development of a missile with France and the purchase of an off-the-shelf missile from American suppliers because Boeing continued to research into the development of the SRAM-T for Britain.

Taormina Compromise Maintains Nuclear Forces

Although the meeting of the NPG in Taormina, Sicily, on 17-18 October 1991 was supposed to resolve the discussion about the future of tactical weapons in Europe, a decision about the TASM was not reached. At the meeting the British Defence Minister King continued to insist on deploying the TASM. King predictably clashed with the German Foreign Minister Genscher who in turn reiterated his demands for the dismantling of all short-range nuclear weapons in Europe, including air-launched missiles. While the Belgian, Dutch and Danish representatives supported Genscher's position, the US delegation came once again to the rescue of Britain. In their Final Communiqué, the members of the NPG went little beyond the reductions that had been achieved between the two superpowers. In addition, NATO members declared that they would cut their stockpile of sub-strategic weapons, including the British WE-177 free-fall nuclear bombs, by 80 per cent. However, in a paragraph born out of British insistence, the NPG members declared that NATO would 'continue to base effective and up-to-date sub-strategic nuclear forces in Europe'. These would 'consist solely of dual-capable aircraft, with continued widespread participation in nuclear roles and peacetime basing by Allies'.

The 'New Strategic Concept' of NATO which was announced at the summit in Rome 7-8 November 1991 essentially reiterated the Taormina compromise. With regard to NATO's nuclear force structure it stated that, due to the changed nature of the threats faced by NATO and the recent successes in nuclear disarmament, sub-strategic weapons could be significantly reduced. However, NATO would 'maintain adequate sub-strategic forces based in Europe, which will provide an essential link with strategic nuclear forces, reinforcing the trans-Atlantic link. These [would] consist solely of dual capable aircraft that could, if necessary, be supplemented by offshore systems'. Tellingly the phrase that NATO members were expected to base dual-capable aircraft in their countries was omitted. *De facto*, the TASM had been eliminated from NATO's defence doctrine. Whether NATO member states were willing to develop and deploy the missile would rest on national decisions. The US government immediately drew its conclusions from the Alliance decision. Speaking to the Senate Armed Services Committee, Under-Secretary of Defence Paul Wolfowitz announced that neither the American government nor NATO planned to develop an alternative to the SRAM-T as tactical air-to-surface missile.

Politicians and Military Split over TASM Future

Although the second stage of the TASM debate had seen the expansion of the international opposition to the missile from Europe to the US, and the abandonment of the programme by NATO, the British government persisted in its determination to go ahead with its TASM. While their limited exposure to international pressure meant that the Cabinet and Conservative MPs were able to maintain their support for the TASM, officials in the Ministry of Defence as the main target of the international protests over the missile was increasingly divided. However, during the winter of 1991-92, the British government remained firmly committed to TASM despite the cancellation of the requirement within NATO and the US. Defence Secretary King defended the development of a sub-strategic deterrent in the House of Commons. British Ministers emphasized in particular the dangers associated with nuclear proliferation after the breakup of the Soviet Union. Since the international support for the missile was lost, it became crucial for the government and MoD officials to prevent increasing pressure for the cancellation of the TASM among the British actors to whom they were linked in the multilevel European foreign policy network. As long as the domestic actors within the network unanimously supported the missile, the government was able to resist the international calls for the abolition of the missile. However, three sets of actors were increasingly susceptible to a change of preference regarding the TASM because their linkages within the network exposed them to considerable transgovernmental pressure or because they were already divided over the issue: MoD staff, the new Defence Secretary Rifkind and the British public. MoD officials, in particular, had close direct relations with their colleagues in the US Pentagon and the other defence departments within NATO. Since their counterparts in NATO had collectively abandoned the missile, the pressure on the British MoD staff to follow suit had increased from 38 to 48 per cent.[18] The transgovernmental pressure was compounded by the financial problems at the

MoD, which had suffered from a series of budget cuts since the late 1980s. To reduce spending on the TASM, RAF officers suggested limiting the range of the missile.

General Election Postpones Decision on Nuclear Missile

However, during the campaign for the upcoming British general elections in April 1992, the government again decided to postpone a decision regarding the missile. The position of the electorate on the question of nuclear weapons was still unclear, and the government preferred to keep the contentious project out of the public debate. Labour and Liberal Democratic MPs shared this sentiment and refrained from using the media to increase the public attention to the issue. Although the members of the two opposition parties publicly expressed their intentions to cancel the purchase of a TASM, these statements remained very low-key. Specifically, the Labour and Liberal Democrats leadership had agreed with each other not to purchase a TASM in the case of a hung parliament. However, the members of the two opposition parties wanted to avoid an anti-nuclear image. They feared that Conservative ministers and MPs would be able to ridicule such a position as unrealistic.

At the heart of their reluctance was the continued insecurity among the members of the government and the opposition about the preferences of the electorate. On the one hand, the government believed that the British public supported the country's nuclear forces. Moreover, the Trident nuclear submarine was associated with jobs in the marginal constituency of Barrow-in-Furness and the Aldermaston Atomic Weapons Establishment. On the other hand, senior officials believed that the electorate was increasingly doubtful of the alleged Russian threat. Nevertheless, Labour leaders welcomed the achievements in the disarmament negotiations between the two superpowers in its party programme 'Agenda for Change' and reiterated their long-term goal of the 'total elimination of nuclear weapons worldwide' (Labour Party, 1992: 22). In Parliament, Labour and Liberal Democratic MPs urged the government at least to reduce its nuclear deterrent by cutting the purchase of a fourth Trident submarine and cancelling the TASM. However, because of the blocking majority of the Conservatives in the House of Commons, the combined pressure from opposition MPs of 46 per cent was unsuccessful in bringing about a change of policy.[19]

MoD Favours Trident Missile Alternative

In the MoD the controversy over the form of Britain's nuclear deterrent intensified during spring 1992. Since the withdrawal of American funds for the missile had increased the development cost of the TASM, MoD officials felt increasingly forced to choose between a fourth Trident and the TASM. The cancellation of Trident would enable the ministry to spend more on the missile, perhaps to increase its range. Sir Michael Quinlan, Permanent Secretary at the MoD and Mr Clark supported this idea. Conversely, the abolition of the TASM would enable MoD staff to redirect the funds to the army and other projects which had been severely affected by previous budgetary cutbacks. By May 1992 first signs indicated that MoD officials had decided against the TASM. They tasked Royal Navy planners with the assessment of a sub-strategic role for Trident, which would allow the TASM to be cancelled. Moreover, the Defence White Paper, published on 7 July 1992, failed for the first time to mention the TASM. Conversely, it stated that the government was 'studying possible replacements' (Statement on the Defence Estimates 1992, 1992: 28) for the WE-177 free fall nuclear bomb - a clear reference to a sub-strategic role for Trident.

As the policy change within the US and NATO had raised the pressure on the staff in the British Ministry of Defence, the preference change among MoD officials in turn increased the opposition to the TASM within the British administration. Specifically, the actors who were directly linked to MoD officials, such as Secretary of Defence Malcolm Rifkind and Prime Minister John Major, were exposed to rising pressure to abandon the TASM. With the military making the case for a sub-strategic role for Trident, the number of actors who advocated a cancellation of the missile increased from 34 to 38 per cent among those who were linked to Rifkind[20] and from 26 to 28 per cent of John Major's linkages.[21] Since the House of Commons, which was subject to the highest pressure among the British decision units at 52 per cent,[22] was prevented from a policy change by the blocking majority of the Conservative MPs, the Secretary of Defence was the most likely to abandon the TASM programme.

Conservatives Rally Behind TASM

At this point, however, the impending abolition of the TASM project triggered countervailing action from Conservative MPs in the Commons and the members of the Defence Committee. Both were insulated from the pressure of the international actors who advocated the abolition of the nuclear missile project. In fact, among the critics of the TASM only MoD officials were able to exert influence over Conservative MPs and the members of the Defence Committee, as they were dependent on the MoD staff for expert information. However, MoD officials amounted to only 8 per cent and 13 per cent respectively of the actors to whom Conservative MPs and the Defence

Committee members were linked in the European foreign policy network.[23] Since most other actors with whom they had close relations remained supportive of the missile, the Conservative MPs and committee members could afford to disregard the pressure from MoD staff for the cancellation. Conversely, alerted by the news that defence ministers were planning to abandon the development of the TASM, the Conservative members of the Commons Defence Committee demanded an inquiry into the issue. The chairman of the committee, Sir Nicholas Bonsor, commented: 'I don't see why this nuclear capability is no longer needed, when one looks to the Middle East'.[24]

Indeed, after the government had argued for two years that the TASM was essential to British defence because of new threats from rogue states like Iraq, cabinet ministers were at a loss to explain why they now wanted to cancel the project. Since Conservative MPs remained staunchly committed to the TASM, the government chose to continue with the development for the time being. According to *Defense Daily*, the British government awarded the three contenders for the missile follow-on contracts, Martin Marietta, Boeing and Aerospatiale, in September 1992 each 1.6 million dollars for a pre-project definition study into a tactical air-to-surface missile, including the development of a prototype. The claim that, if TASM was made a NATO requirement in 1995, Britain would be able to profit from sales to other NATO members once the TASM design was completed was scarcely convincing, however.

Defence Secretary Rifkind for Cancellation of TASM

By autumn, new international problems emerged with regard to the American contenders for the British missile. A nuclear test ban installed by the US Congress on 1 October 1992 put the British TASM project further under strain because it prevented British scientists from testing the new warhead at US sites. Although the missile programme was increasingly unlikely, the British government urged the US to resume its nuclear tests as late as July 1993. Similar problems challenged the French ASLP. After the Russian Defence Minister Pavel Grachev announced that his government would extend its temporary nuclear test ban until at least mid-1993, a French Defence Ministry spokesman stated that it was considering extending its own one-year moratorium which was to end. Although the French government assured that the test ban would not affect its nuclear capabilities, the ASLP appeared to run into further problems as the French government, too, was pressed to reduce its defence spending.

In October 1992, Defence Secretary Malcolm Rifkind withdrew his support for the missile project. The minister not only had to take into account pressure from his staff for a cancellation of the missile, but also the international opposition from his NATO colleagues to the TASM, which together amounted to 38 per cent of his contacts in the network. Speaking to journalists, Rifkind acknowledged that Britain's defence commitments might have to be reduced. He intimated that the cuts could include the TASM. The Defence Minister warned the Cabinet, however, that further cuts would force a reconsideration of British strategy. At the meeting of NATO defence ministers in Gleneagles, Rifkind stated that the review into the options for a British TASM would be completed by early 1993. The design of a nuclear warhead for the TASM had almost been finalized. The Defence Ministry had earmarked 1.5 million pounds for 1993 to determine a missile platform for its TASM programme which was to be decided soon.

British MPs Under Pressure

Although the fact that Defence Secretary Rifkind was prepared to cut the TASM slightly increased the pressure on Parliament to accept the cancellation of the missile, the comparative insulation of Conservative MPs and the members of the Commons Defence Committee from international actors limited the pressure on both. Thus, while Defence Committee members accepted in their First Report of 1992 that 'financial considerations will quite properly play a part' in the decision over the future of the TASM project, they warned the government that the 'risks inherent in any proposal whereby strategic and sub-strategic deterrents are dependent upon the same launch platform', namely the Trident submarines, had to be addressed (Defence Committee, 1992: x). In February, Labour and Liberal Democrat MPs used the increasing differences within the government to propose to the House of Commons the abolition of the TASM in order to save three billion pounds for defence cuts which had been requested by Treasury staff. However, since the opponents of the missile could only exert indirect pressure on the Conservatives via Parliament, the Conservative MPs were able to maintain their resistance. In the Commons, the Conservative majority could easily block the proposed cancellation.

The preference changes of the British MoD and Secretary of Defence Malcolm Rifkind during the summer of 1992 extended the transgovernmental pressure for the cancellation of the TASM programme among the British actors in the multilevel European foreign policy network. However, the government hesitated in cutting the TASM. In particular, the blocking position of Conservative MPs in the Commons meant that the government had to convince its parliamentary party first before it could risk a vote on a defence budget which incorporated the abolition of the TASM programme. The struggle of the

British administration to build a consensus among domestic actors for the cancellation of the TASM extended over another year from October 1992 to October 1993.

TASM Cancellation Pushed through Parliament

By the beginning of 1992 RAF staff were resigned to losing their role in Britain's nuclear defence. By giving up the TASM, Air Force officers could reallocate resources to retain the Eurofighter Aircraft programme. The 1993 Defence White Paper 'Defending Our Future' published in July was expected to give the final blow to the TASM project. However, the defence review merely stated that the decision would be announced in due course. In a concession to Conservative opposition, the government's White Paper emphasized the continued need for sub-strategic nuclear weapons. In fact, the White Paper mentioned the 'Provision of an Effective Independent Strategic and Substrategic Nuclear Capability' first among Britain's military tasks (Statement on the Defence Estimates 1993, 1993: 14).

By the time of the publication of the White Paper, however, Defence Secretary Rifkind and MoD officials were making progress in convincing Conservative MPs that a tactical missile for Trident could substitute for a TASM. In particular, the government had focussed its efforts rationally on the members of the Commons Defence Committee who were, at 19 per cent,[25] under marginally higher collective pressure within the network than the Conservatives at 17 per cent.[26] After extensive briefings from the MoD staff, the members of the Defence Committee changed their position on the TASM. In June, the Defence Committee announced that there were no reasons why Trident could not take the sub-strategic role originally envisaged for the TASM.

For the following development of a winning coalition in favour of a cancellation of the TASM, the preference change among the members of the Defence Committee proved critical. In particular, the support for the abolition of the TASM programme by the committee members increased the pressure on Conservative MPs, among whom the committee members had particular authority on matters of defence, to 25 per cent of their links in the network.[27] Furthermore, due to the preference change of the committee members, the pressure on the House of Commons as a collective decision unit rose to 59 per cent.[28] The considerable support for the cancellation of the TASM in Parliament meant that it became increasingly politically unviable for the Conservative majority to block the abolition.

Following the approval of committee members of the abolition of the TASM project, Defence Secretary Rifkind raised the issue in the Cabinet on 30

September 1993. Although a small number of cabinet ministers were critical of the severe cuts in the defence budget, none challenged the 'technical' choice between the TASM and a sub-strategic Trident missile. Moreover, when the Cabinet announced its decision to cancel the TASM in Parliament, Conservative MPs acceded without major protests. With the Cabinet's official change in policy an overwhelming 55 per cent of the actors to whom the Conservative MPs were connected in the network supported the decision, thus raising the political stakes of a rejection.[29] Their acceptance was made easier by a preceding public showdown over the defence budget between the Chancellor of Exchequer Kenneth Clarke and Defence Secretary Malcolm Rifkind. To the Conservative MPs and the media, the government presented the TASM as the sacrificial lamb which not only helped to prevent job losses in the armed forces of more than 20 per cent, but also the cancellation of the European Fighter Aircraft. The budget argument was preferable to admitting to the international pressures which had influenced the government's decision. It maintained the image that British nuclear decision-making remained a national preserve. Although Conservative backbenchers continued to express fears that the abolition of a nuclear role for the RAF would undermine Britain's nuclear deterrent, the threat of further redundancies among the armed services was a more serious concern to them. When Defence Secretary Rifkind announced in the House of Commons on 18 October 1993 that Britain was backing out of the TASM programme, the Conservative Party did not challenge the decision.

Assessment of the Hypotheses

As in the previous case studies, five indicators are used to evaluate the hypotheses proposed by multilevel network theory in this chapter: the frequency of preference changes with rising degrees of pressure, the distribution of pressure in the four behavioural categories no change of preferences, unclear or undecided preferences, change of preferences and blocked preference changes, the average pressure in each behavioural category and the timing of the preference changes. These measures not only help to assess to what degree the hypotheses were corroborated, but also suggest new inductive propositions concerning the probabilistic causal relationship between pressure and preference changes.

The first measure which gives the proportion of preference changes in relation to other forms of behaviour at different degrees of pressure generally confirms the first hypothesis. Thus, the results in Figure 5.1 show that the relative number of preference reversals increased virtually consistently with rising degrees of pressure. While at pressures of 10-15 per cent only 20 per cent of the actors changed their policy preferences, at 30-35 per cent it were more than 30 per cent, at 40-45 more than 50 per cent and at 55-60 per cent of pressure 100 per cent.

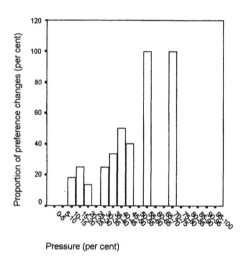

Figure 5.1   Proportion of preference changes

The utility of the first hypothesis is further corroborated by the second indicator which shows the distribution of the four types of behaviour across different degrees of pressure. Specifically, the findings displayed in Figure 5.2

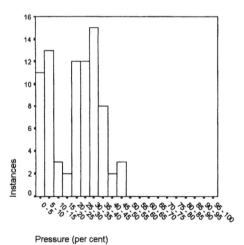

Figure 5.2   No change of preferences

demonstrate that the number of instances in which actors were able to maintain their original policy preferences falls from its peak at pressures between 30-35 per cent to zero at pressures above 50 per cent. The inductive proposition of a threshold at pressures of about 50 per cent is again supported by the quantitative findings. As in the case of air strikes in Bosnia, none of the actors was able to resist degrees of pressure higher than 50 per cent.

The distribution of actors or instances whose preferences were unclear or who were undecided in Figure 5.3 is less insightful because of the few number of instances of such behaviour in this case study. In so far as general conclusions might be drawn from the distribution, it appears conform with the inductive proposition that actors experience a phase of reorientation at intermediate degrees of pressure which had been suggested in the previous chapters. However, in the case of the British TASM, none of the actors seems to have used unclear preferences to resist high degrees of pressure.

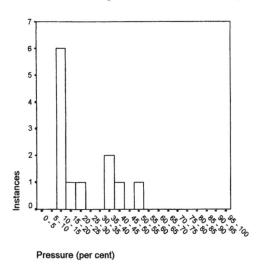

Pressure (per cent)

Figure 5.3   Unclear/undecided preferences

The instances in which actors changed their preferences regarding the TASM project in Figure 5.4, again confirms the first hypothesis of multilevel network theory. Thus the figure shows that the number of preference changes increases significantly at higher degrees of pressure. While the number of instances in which actors maintained their preferences in Figure 5.2 peaks at a range between 30-35 per cent, the number of cases in which actors changed their position on the TASM in Figure 5.4 shows the highest occurrence between 30 and 40 per cent. The findings thus illustrate clearly that actors typically begin to change their preferences when they become subject to pressure from about 30 per cent of the actors to whom there are linked in the network. Indeed, there are only few preference changes among actors who are subject to pressure from less than 30 per cent of their network linkages. In addition there are only very few instances in which actors changed their preferences when subject to pressures above 50 per cent which lends further support for the suggestion of a threshold at this level of pressure.

Blocking behaviour which is explained by the second hypothesis was a possible strategy for only a very limited number of actors in the multilevel European foreign policy network regarding the decision over the TASM

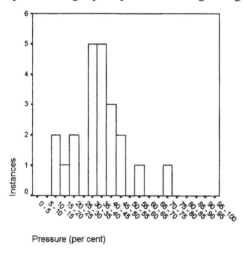

Figure 5.4 Change of preferences

programme. Since the issue concerned nuclear defence policy, most international organizations did not have any authority over the decisions of the key actors - Britain, France, or the US. The only exception was the Nuclear Planning Group of NATO. However, its decision-making capacity applied merely to the nuclear defence doctrine of the Atlantic Alliance as a collective organization to which the member states designated specific weapon systems. Whether and how the three Western nuclear powers decided to maintain their nuclear defence capabilities was under the exclusive authority of their governments. It is, therefore, questionable if the British government used its veto in the NPG. It could have been possible since formally all Alliance members hold a veto in the collective decision-making process. However, the fact that a veto was not necessary to maintain Britain's freedom of action on the issue and the observation that NATO changed its defence requirement immediately after the US had withdrawn its support for the development of the SRAM-T, speak against such speculation. If Britain had exercised a blocking strategy, it should at least have delayed the NATO decision. In fact, there was no evidence of a veto in the primary sources.

The only blocking behaviour which is shown in Figure 5.5 was found by the members of the Conservative Parliamentary Party who opposed a cancellation of the TASM project within the House of Commons. At degrees between 45 and 60 per cent, it generally appears to support the second

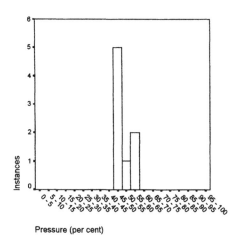

Pressure (per cent)

Figure 5.5   Blocked preference changes

hypothesis which suggests that it enables the affected collective decision-making bodies to resist somewhat higher degrees of pressure than in cases where where none of the members blocks a decision. However, the final control of the House of Commons over the decision to abandon the TASM programme was rather restricted. The approval of the House of Commons to nuclear policy could only be given to the Defence White Paper as a whole, not to single projects. Since a rejection of the White Paper would be synonymous with a vote of no-confidence, the stakes against such action were prohibitively high. Nevertheless, in the decision regarding the future of the TASM programme, the Cabinet went through great efforts to receive the backing of its Parliamentary Party. For nearly three years, from summer 1990 to summer 1993, the support of the Conservative Parliamentary Party was crucial in order to fend off the collective pressure of the opposition parties for the cancellation of the TASM. During this period, the Conservative majority in the House of Commons successfully vetoed a rejection of the TASM programme. Even after key cabinet members had changed their preference with regard to the missile, the blocking majority remained in place - much to the detriment of the government which now increased their pressure on the reluctant Conservative

MPs. At the maximum, the Commons were able to withstand a pressure of 72 per cent of the actors to whom it had direct links in the network. Due to the Common's nature as a collective decision-making body, however, the crucial degree of pressure was that on the Conservative MPs who had the parliamentary majority. Incidentally, the pressure on the Conservative MPs amounted to only 50 per cent and thus explains the ability of the Commons as a collective body to resist the great amount of pressure placed on it.

Table 5.1 Descriptive statistics

Preference changes	Number of instances	Range of pressure	Minimum pressure	Maximum pressure	Average pressure
No change (NC)	81	48%	0%	48%	22%
Unclear or undecided (U)	12	37%	14%	52%	25%
Change (C)	22	59%	13%	72%	36%
Blocked (B)	8	12%	46%	59%	50%

The third indicator, namely the average degrees of pressure given in Table 5.1, clearly supports the two hypotheses. Notably, there are considerable differences among the average pressures in each of the four behavioural categories 'no change', 'unclear or undecided', 'change' and 'blocked' preferences. The average pressure for a change of their foreign policy preferences which actors were able to resist was, at 22 per cent, significantly lower than the average pressure at which actors succumbed to this pressure at 36 per cent.

The mean of pressure at which network actors were divided over their preference or unclear was also below the mean of pressure at which the modified their original policy preferences. More clearly than the distribution of this behavioural category across the full range of pressures between zero and 100 per cent, the average thus supports the inductive proposition that actors pass through a phase of reorientation expressed by unclear preferences before they adopt a new policy preference.

Since the strategy of the Conservative Parliamentary Party was the only instance of blocking behaviour, the average degree of pressure which Parliament was able to withstand is less insightful than in the previous case study. However, at 50 per cent it generally supports the second hypothesis which proposes that collective decision-units are able resist higher degrees of pressure than other actors if one member exerts a veto or if a majority blocks a decision.

The final indicator, the correlation between the degree of pressure exerted on network actors and the timing of their preference changes, is supported even more strongly than in the other case studies. Out of fifteen actors who changed their preference during the decision-making process, thirteen modified their position immediately after the pressure on them had increased, including the members of the US Congress, the US Secretary of Defence, staff from the US State Department, US President Bush, the NATO Council of Ministers, NATO military, officials from the British Ministry of Defence, the British Defence Secretary, the British Foreign Secretary, the British Cabinet, the members of the House of Commons Defence Committee, the members of the House of Commons and the members of the Conservative Parliamentary Party. The US military and the US Secretary of State abandoned their support for the TASM project in the second phase after the pressure on them had been raised. None of the actors were able to withstand an increase in the degree of external pressure for more than two consecutive periods.

Conclusion

From the findings presented in this chapter, three conclusions can be drawn concerning the new insights provided by multilevel network theory into the role of the multilevel European foreign policy network in the making of national defence policy.

The most fundamental observation is that despite the efforts of a number of European countries, in particular Britain and France, to maintain the impression that their national defence policy is determined exclusively by domestic considerations, transnational and international actors had a considerable impact on the relevant decision-making processes. Thus the preceding analysis has demonstrated that the British redefinition of its nuclear requirements was, if not brought about, at least significantly influenced by Britain's NATO allies who pressed for the abolition of most nuclear weapons in Europe in the aftermath of 1990. Specifically, this chapter reveals the national, transnational and international extent of the coalition which emerged in opposition to the TASM programme. It shows how already in the initial stages of the decision-making process ministers and officials from most NATO member states exerted transgovernmental pressure on their British and American counterparts to cancel the programme. This pressure explains why, when the US Congress threatened to withdraw crucial funds for a TASM by Boeing, the missile lost the support of the American administration. Moreover, multilevel network analysis illustrates how the US and the European governments which opposed the TASM formed an international coalition within NATO which led the North Atlantic Council to cut its requirement for TASMs by December 1991 and raise the international pressure on the British government. Indeed, the preceding analysis illustrates that the subsequent preferences change among staff of the British MoD and, eventually, the British government could to a large degree be attributed to these pressures.

The second conclusion concerning the making of national defence policy in Europe is that, even in the area of national security where institutional integration has been limited, the British government was not able to act as a gatekeeper. Despite the importance of the boundary and bridging roles of ministers and civil servants between the national and international arenas, the gatekeeper concept was specifically challenged in two ways. First, transnational pressure, such as that from US Congress members on British parliamentarians, was able to circumvent the administration. Second, the administration itself was divided and, as a consequence, could not prevent transnational influence. As in the preceding case study, the control of government ministers over Britain's national defence policy was undermined by the ability and willingness of civil servants to forge a transgovernmental coalition with their colleagues in other NATO countries and the pressure which this coalition exerted on the politicians.

Finally, it can be observed that a stronger focus of network theory on the process of decision-making as proposed in this book helps to explain the timing of the British decision to cancel its TASM project nearly three years after the end of the Cold War. In particular, the analysis of the national and international interactions preceding the abolition of the TASM programme explain why the British administration invested 1.6 million dollars into the development of the missile as late as September 1992. According to multilevel network theory the prolonged deferral of a decision regarding the future of the TASM was due to the lack of direct influence over the government among the actors who advocated an abolition of the missile. In particular, the Cabinet and Parliament which held the ultimate decision-making authority had no direct relations with other national administrations and were insulated from the international opposition to the TASM which had emerged within NATO. As a consequence, the pressure for the abolition of the weapon had to build up more gradually through indirect linkages. As this chapter illustrated, the preference change among officials in the British MoD who were linked transnationally to their NATO colleagues and nationally to British politicians was essential for this build-up. By forming a transgovernmental coalition in favour of the cancellation of the TASM, the British military helped to extend the international pressure among domestic actors in Britain. It thereby fostered the formation of a winning coalition between cabinet ministers and the members of the Commons Defence Committee which eventually overcame the objections of Conservative MPs to the cancellation of the missile in the House of Commons.

Interestingly, the findings again suggests that a policy change only came about after a majority of national as well as international actors had come to favour the abolition of its TASM programme. Although the case study illustrates that international pressures played a considerable role in bringing about the policy change of the British government, it also shows that the government resisted these pressures until it had gained the support of a range of domestic actors, in particular of the Conservative MPs, the House of Commons and the members of the Defence Committee, for the cancellation of the missile project.

Notes

1. Specifically, the defence ministers and staff from the four countries accounted for eight out of 29 actors to whom the British Secretary of Defence was linked, raising the pressure on him to P_1 [DS] = 8/29 = 28%.
2. In the North Atlantic Council the seven member states were represented through their heads of state, their foreign or defence ministers, thus accounting for E=3x7=21 links in the network and raising the pressure to P_1 [Nato CM] = 21/51 = 41%.
3. Thus, the pressure from their counterparts in seven NATO member states accounted for 37 per cent of the Pentagon staffs' linkages in the network, i.e. P_1 [US-pen] = 7/19 = 37%.
4. The number of linkages of the US President was even higher than those of the US Defence Secretary raising the pressure to 25 per cent, i.e. P_1 [US-Pre] = 8/32 = 25%.
5. The number of linkages of the US President was even higher than those of the US Defence Secretary raising the pressure to 25 per cent, i.e. P_1 [US-Pre] = 8/32 = 25%.
6. The pressure from the seven NATO countries accounted respectively for P_1 [Mod] = 7/21 = 38%, P_1 [FS] = 8/36 = 22% and P_1 [PM] = 8/39 = 21%.
7. Quoted in Martin Walker, 'Soviet generals may be behind nuclear stance', *Guardian*, 7 June 1990.
8. With 251 Labour and Liberal Democratic MPs using the 16 linkages of Parliament to oppose the TASM and exerting pressure on the remaining 399 MPs, E=(16x251)+(2x399) =4814 and P_1 [Par] = 4814/10400 = 46%.
9. Among the fifteen linkages of the Defence Committee members in the multilevel European foreign policy network, only the Labour MPs opposed the TASM, raising the pressure on them to P_1 [Com] = 1/16 = 6%.
10. Andrew McEwen, 'Britain 'poised to pick France' as partner in nuclear missile project', *Times*, 24 October 1990.
11. Christy Campbell, 'Nuclear role for Tornado crews', *Sunday Telegraph*, 28 April 1991.
12. Due to the preference change among Pentagon officials, the number of actors who favoured the cancellation of the missile increased from P_2 [US-SD] = 9/24 = 38% to P_3 [US-SD] = 10/24 = 42% among those who were directly linked to Defence Secretary Cheney.
13. The Pentagon staff also increased the pressure on the US President from P_2 [US-Pre] =9/32 = 28% to P_3 [US-Pre] = 10/32 = 31% of his contacts in the network.
14. Nick Cook, 'The tactical missile debate that refuses to lie down', *Jane's Defence Weekly* 16:16, 19 October 1991, p.708.

15. Among the linkages of NATO's bureaucratic organization, officials from the foreign and defence departments of US, Germany, the Netherlands, Belgium, Italy, Denmark, Norway and Iceland favoured the cancellation of the TASM, accounting for P_4 [Nato-Org] = 16/29 = 55%.

16. At this time 24 out of 51 actors linked to the North Atlantic Council opposed the TASM, namely the representatives of Germany, the Netherlands, Belgium, Italy, Denmark, Norway, Iceland and the US, accounting for P_4 [Nato-CM] = 24/51 = 47%.

17. Marc Rogers, 'NATO relived as SRAM-T is cut', *Jane's Defence Weekly* 16:15, 12 October 1991, p.643.

18. The preference changes among the Pentagon staff and NATO military increased the pressure on MoD officials from P_2 [Mod] = 8/21 = 38% to P_5 [Mod] = 9/21 = 43% of their linkages in the network.

19. P_5 [Par] = 4814/10400 = 45%.

20. The preference change among the MoD staff increased the pressure on Defence Secretary Rifkind from P_5 [DS] = 10/29 = 34% to P_6 [DS] = 11/29 = 38% of the actors to whom he was linked in the network.

21. MoD officials further increased the number of actors who supported a cancellation of the British TASM from P_5 [PM] = 10/39 = 26% to P_6 [PM] = 11/39 = 28% among the network linkages of Prime Minister Major.

22. Due to changes in the composition of the House of Commons following the April 1992 General Election, P_6 [Par] = 5376/10416= 48%.

23. The preference change among MoD officials raising the pressure on the members of the Defence Committee to P_6 [Com] = 2/16 = 13%, while the pressure on the Conservatives remained constant.

24. Colin Brown, 'Tory MP's oppose plans to scrap new missile', *Independent*, 13 August 1992.

25. The preference change of Rifkind had increased the number of actors who opposed the TASM among the linkages of the Defence Committee members to three, accounting for P_7 [Com] = 3/15 = 20%.

26. The pressure on the Conservatives remained at P_7 [con] = 2/12 = 17%.

27. Specifically, three out of eleven actors to whom the Conservatives were linked in the network supported the cancellation of the TASM, namely Defence Secretary Rifkind, the members of the Defence Committee and US politicians, accounting for P_8 [con] = 3/11 = 27%.

28. P_8 [Par] = 6096/10416 = 59%.

29. The open advocacy of the TASM cancellation by the Cabinet, including Prime Minister Major and Foreign Secretary Hurd increased the number of the actors who exerted pressure on the Conservatives by three to P_9 [con] = 6/12 = 50%.

6   Conclusion

This book began with the observation that the study of contemporary European foreign policy increasingly requires the combination of multiple levels of analysis. As decision-making has come to involve a broad range of political, social and economic actors in the national and international arena, the division between theories of international relations and foreign policy analysis has been criticized for inhibiting more comprehensive explanations of foreign policy processes and outcomes. A range of multilevel models, such as transnationalism, the two-level game and network models, offer insights into the integrated policy decision-making processes in this area. This book has sought to demonstrate that the network approach is particularly well suited for modelling the increasing multiplicity, diversity and interdependence of actors across the national, transnational and international levels of analysis. However, it has pointed out that most existing network models fail to address sufficiently the theoretical implications of integrating all three levels of analysis. Moreover, few network models propose testable hypotheses regarding decision-making processes as an intermediate variable between structure and outcomes. This book has proposed a multilevel network theory which seeks to address both criticisms. Specifically, multilevel network theory suggests a new way in which rational choice theory may be used to explain how public and private actors are able to influence the foreign policy decision-making process and its outcome. In addition, this book sought to provide new empirical insights into the question who makes European foreign policy and how.

By returning to the theoretical and empirical concerns outlined in the introduction, this chapter seeks to provide an overall assessment of multilevel network theory and its utility for explaining contemporary European foreign policy making within and beyond the European Union. The first section analyses to what degree the findings from the three case studies of foreign policy making in the EU, the transatlantic community and the United Kingdom have corroborated the hypotheses proposed by multilevel network theory. The second section revisits the three theoretical approaches to multilevel decision-making outlined in the introduction of this book and discusses in how far multilevel network theory provides new and different insights into European foreign policy decision-making. And the final section examines the implications of the empirical findings presented in this book for the question who defines European foreign policies.

A Cross-Case Evaluation of Multilevel Network Theory

The preceding chapters have sought to evaluate the validity of the hypotheses proposed by multilevel network theory for the explanation of European foreign policy decision-making in three distinct contexts: the European Union, the transatlantic community and national security. This section examines to what degree the two hypotheses were corroborated across the three case studies, where they were challenged and whether the empirical findings suggest more specific or additional hypotheses concerning the decision-making process and policy preference changes among foreign policy actors across multiple levels analysis.

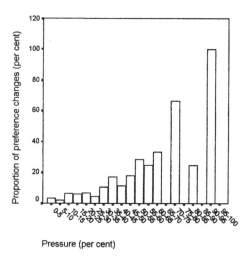

Pressure (per cent)

Figure 6.1   Proportion of preference changes

According to the first hypothesis the higher the degree of pressure (P=E/L), i.e. number of directly related actors exerting pressure (E) on an actor X out of all how have power over him or her (L), the more likely is actor X to change his or her policy preference. It follows that the proportion of preference changes among the possible forms of action, which include no change of preferences, unclear or undecided preferences or a modification of preferences, should increase with rising degrees of pressure.

While the single case studies have already confirmed this proposition, the cumulative empirical evidence from the three cases illustrates more clearly the degree to which this hypothesis is corroborated by the findings. As Figure 6.1 shows, the relative frequency of a preference change rises almost consistently when the pressure on any actor of the network increases.

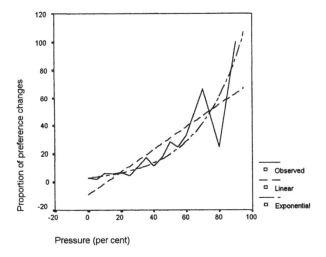

Pressure (per cent)

Figure 6.2   Proportion of preference changes - curve fit

However, attempts to fit a curve to the graph representing these findings indicate that the first hypothesis can be refined. While the original first hypothesis does not specify the scale of the increase in the likelihood of a preference change with rising degrees of pressure, i.e. the steepness of the distribution, the empirical evidence across all cases suggest that this increase might not be linear. Rather it appears, as Figure 6.2 demonstrates, that the probability of an actor changing his or her policy preference rises exponentially with increasing degrees of pressure. A modified first hypothesis can, therefore, be stated as:

> The probability of an actor changing his or her policy preference increases **exponentially** with rising degrees of pressure (P=E/L), i.e. number of actors exerting pressure (E) on a single actor X out of all actors who have power over him or her (L).

Additional new insights into the likelihood of a policy preference change and the behaviour of foreign policy actors in multilevel networks are provided by the analysis of each of the four behavioural categories. Especially the distribution of the instances in which actors maintained their original preferences in Figure 6.3 offers further understanding as to when actors change their policy preferences during the decision-making process and how this is related to the pressure which actors exert on each other. Specifically, the cumulative analysis shows that most actors consistently resisted pressures to change their policy preferences if less than 45-50 per cent of the actors who had power over them advocated a different policy. However, once the degree of pressure on any actor exceeded 55 per cent, they began to respond either by displaying unclear preferences or changing them.

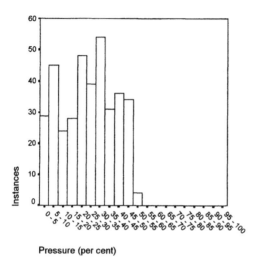

Pressure (per cent)

Figure 6.3   No change of preferences

The empirical findings thus suggest the additional proposition of a threshold. Remarkably, this threshold lies almost exactly at a pressure of 50 per cent which characterizes a situation in which an actor is exposed to pressure from half of the actors on which he or she is linked within the network. Incidentally it conforms with the notion that in democracies a policy is or should be determined by the preference of the majority. However, the findings derived from the previous case studies suggest that the majority not only influences the policy preferences of politicians or parliaments, but in fact all national, transnational and international actors within a network. Who is part of the majority view differs for each actor or decision-making unit because every member in the network has a different constituency, i.e. is linked to a different group of actors. It is the policy preferences of this group which matter to each actor.

Although multilevel network theory did not predict the existence of a threshold, it generally conforms with the premises and propositions of the model and thus can be incorporated into the approach. On the basis of the empirical findings, the new additional inductive hypothesis can be stated as follows:

Actors are unlikely to maintain their original preferences at a degree of pressure (P=E/L) higher than 50 per cent if their preference change is not blocked. They will either change their preferences, or appear unclear or undecided over their preferred policies.

The existence of a threshold implies that the degree of pressure is a sufficient, but not necessary explanation for changing preferences. In addition to the pressure from other actors within their network, each actor apparently also considers other variables. They can help to explain whether an actor modifies his or her preference at pressures lower than 50 per cent. The nature of these variables and the degree to which they influence policy changes can only be established in further studies. However, the case studies indicate that once the pressure reaches the threshold level, alternative considerations recede into the background of an actor's calculations. As multilevel network theory suggests, rational actors should not and, as the case studies confirm, do not ignore the policy preferences of the majority of actors to whom they are linked in their network, regardless of their other motives or the strength of their convictions.

The distribution of pressure in the category in which actors changed their policy preferences in Figure 6.4 provides further evidence for the above argument. While a small number of actors reversed their policy position in the absence of significant pressures and others were able to withstand strong external demands, the number of preference changes peaks between 30 and 50 per cent, i.e. just before the threshold level.

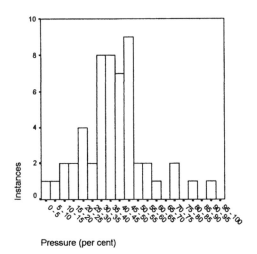

Pressure (per cent)

Figure 6.4 Change of preferences

The few instances in which actors changed their policy preferences at pressures beyond the majority threshold can be explained by differences in the intervals in which the pressure on each individual actor rises. Actors with few linkages within the foreign policy network have larger intervals than actors with a large number of linkages. Sudden, steep increases in pressure are also a result of simultaneous preference changes, often among a set of closely linked actors or towards the end of the research periods when overall pressures on a number of actors have reached high degrees.

The findings from the three cases further suggest that the difference between the behavioural categories 'no change' and 'change' is filled by a third category which has been described as 'unclear or undecided'. The distribution of the pressures in this category, displayed in Figure 6.5, indicates that actors frequently pass through a phase in which their initial conviction falters, but a new policy position has yet to be adopted. As a consequence, an individual or different representatives of a particular role express diverging positions or contradictory preferences. In collective decision-making units which take decisions by vote or consensus, such as parliaments or the ministerial councils of international organizations, this behaviour usually signifies a split among their members.

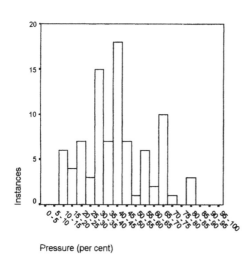

Pressure (per cent)

Figure 6.5   Unclear/undecided preferences

The observation that the number of unclear or undecided positions peaks at pressure of 40-45 per cent , i.e. just before the number of actors who changed their preferences, implies that this phase might precede an eventual preference change. Interesting is the additional observation that in some circumstances unclear preferences seem to enable actors to delay a change in their policy preferences. Thus, in 23 out of 90 instances, unclear preferences appeared to enable actors to withstand pressures beyond the threshold degree of 50 per cent.

According to the second hypothesis proposed by multilevel network theory, the ability to resist network pressure is also enhanced by a veto or blocking position in collective decision-making units, like parliaments or international organizations. That is, a policy preference change in these organizations can be prevented by their members, if they have a veto or if decisions require a (qualified) majority or a consensus. The empirical findings appear to affirm this proposition in the distribution of blocking behaviour over different degrees of pressure in Figure 6.6 which shows that blocking behaviour typically occurred at degrees of pressures between about 40 and 60 per cent and thus at slightly higher degrees of pressure than preference changes.

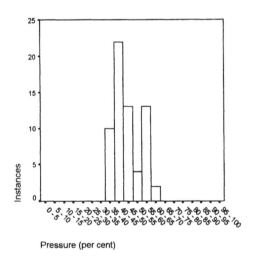

Pressure (per cent)

Figure 6.6   Blocked preference changes

Although it is difficult to make generalizations since blocking strategies were only use in two of the cases, the findings do not challenge the second hypothesis.

What appears interesting, however, is the empirical finding that actors only used a veto strategy if the degree of pressure on their organization increased beyond 35 per cent. This can be explained by the fact that a veto is not necessary to prevent a decision in collective decision-making units unless a considerable number of members support a change in policy. It seems far from incidental that at a starting range of pressures from 35-40 per cent, blocking behaviour falls into the same bracket in which actors on average have a high probability of changing their preferences. Yet considering that decisions in all collective decision-making units required either a majority or a consensus among its members, the degree at which blocking behaviour fails to prevent a preference change appears to be rather low. At about 60 per cent it is only ten per cent above the majority threshold and, thus, far from meeting a qualified majority or, indeed, a consensus.

The observation that vetos failed to prevent changes of policy preferences in collective decision-making units even if less than 60 per cent of the members supported them, illustrates that collective decision-making units are still subject to the first hypothesis. Accordingly, a revised and more specific second hypothesis can be defined as follows:

> Collective decision-making units can resist pressure of **up to 60 per cent** if members use a veto or if a decision requires a (qualified) majority ['veto or blocking strategy'].

The average pressure within each of the four behavioural categories across all cases as shown in Table 6.1 further corroborates the above findings. The first hypothesis is supported by the observation that the mean pressure which actors were able to withstand was, at 26 per cent, significantly lower than the average pressure at which actors changed their policy preferences at 40 per cent. The average pressure at which actors were unclear or undecided lies at 41 per cent only slightly above the average pressure at which actors modified their preferences which further supports the notion that it, like a blocking strategy, also helps actors to resist pressures for a change of their policy preferences. The second hypothesis is also confirmed by the statistical evidence. Thus, at 47 per cent collective decision-making units were able to withstand significantly higher average pressure than unitary role actors without a preference change.

Table 6.1 Descriptive statistics

Preference changes	Number of instances	Range of pressure	Minimum pressure	Maximum pressure	Average pressure
No change (NC)	372	52%	0%	52%	26%
Unclear or undecided (U)	90	67%	14%	81%	41%
Change (C)	53	95%	5%	100%	40%
Blocked (B)	64	28%	35%	64%	47%

Assessed by the standard of empirical corroboration, the hypotheses proposed by multilevel network theory are thus generally confirmed and can even be refined. However, in the final instance, the explanatory value of a theory can only be assessed in comparison with other theoretical models. In order to examine how far the multilevel network theory proposed in this book offers different insights into contemporary European foreign policy decision-making, the following section therefore returns to the three multilevel approaches discussed in the introduction: transnationalism, the two-level game and network models.

Multilevel Network Theory in Comparison

Multilevel network theory suggests several theoretical developments for the analysis of multilevel decision-making. This section examines whether these developments promote a different and fruitful approach to the integration of multiple levels in the study of European foreign policy making. Specifically, it compares the ability of multilevel network theory to analyse a diversity of European foreign policy actors, to explain decision-making processes and to measure the influence of actors at different levels on foreign policy outcomes with that of transnationalism, the two-level game and the British policy network approach.

Diversity of Actors in European Foreign Policy

One of the key differences between multilevel network theory and the three multilevel approaches discussed in the introduction of this book is the range of actors and their relations which are analysed. While transnationalism focuses on transnational relations, two- or three-level games analyse domestic and international demands on international negotiators, and most network models conceptualize domestic levels of analysis, multilevel network theory proposes a possible theoretical integration of all three levels of analysis. Only by doing so can multilevel network theory attempt to examine empirically whether there are significant differences in the way actors at different levels participate in the European foreign policy decision-making process.

In the three test cases, multilevel network theory is thus able to show that national, transnational and international actors not only regularly sought to influence national or international decision-makers, but also each other across levels of analysis. Specifically, the empirical evidence presented in this book refutes the premise of the two- or three-level game according to which governmental representatives at the international level are in the exclusive position to influence the formation of national or transnational coalitions in favour of a particular policy outcome. Although ministers and civil servants were able to exert pressure on a range of actors at the national and international level by means of their boundary role in the network, they were not the only actors which held transnational linkages, nor were they the only actors who influence the formation of coalitions in favour of particular policies. Conversely, the insights gained by multilevel network theory illustrate that all actors sought to use their national, transnational or international relations within the foreign policy network to exert pressure on each other in favour of their most preferred policy. Moreover, the three case studies show that coalitions formed as often within as across levels of analysis.

Explaining Decision-Making Processes

The second difference between multilevel network theory and these three multilevel approaches regards its ability to hypothesize about the process of decision-making. In multilevel network theory this process is understood as a sequence of preference changes which leads to the formation of a winning coalition in favour of a specific policy outcome. By hypothesising about processes, multilevel network theory expands upon existing multilevel approaches in two ways. First, it introduces a concept of agency in the form of rational choice premises into the predominantly structural analysis of transnationalism, two- and three-level games and network models. Second, it seeks to explain rather than describe the foreign policy decision-making process and its outcomes.

  The concept of agency is of particular importance for the explanation of the decision-making process and its outcomes because, as the three case studies indicate, neither the policy preferences nor the coalitions among public and private actors could be regarded as static. Contrary to the assumptions of the two- or three-level game and many network models, actors could and in many instances did change their policy preferences and coalition allegiances during the decision-making process. Indeed, the findings demonstrate that actors even chose whether to engage at all in the decision-making process or whether to remain neutral. Thus, actors joined the public debates regarding a policy only if their preferences were directly affected or if they were under pressure from other network actors to become part of a coalition for or against a policy.

  It follows that the behaviour of actors and the process of decision-making are important elements in the explanation of policy outcomes. Rather than reading a decision off the distribution of preferences among the members of a network, multilevel network theory offers new insights into the decision-making process as an intermediate variable between preferences and outcomes. Specifically, it suggests that outcomes are influenced by a coalition formation process in which actors tried to modify each others' policy preferences. Network structures retain their importance in that they determine which actors have the potential to influence each other. However, this book illustrates that by combining structural analysis with rational choice assumptions, multilevel network theory is able to explain which actors would exert pressure on others and when actors would change their preferences and coalition in response to this pressure.

The formation and changes in the coalitions among network actors ultimately determined whether a winning coalition emerged in favour of a particular policy. According to all four approaches, it is the winning coalition which defines the outcome of the decision-making process. However, while the two- or three-level game and some network models appear to identify the members of this coalition on the basis of similar policy preferences, multilevel network theory advances an alternative view. It contends not only that the members of a coalition have to be linked to each other, but also that, in order to influence policy outcomes, the members of the winning coalition have to be linked to the ultimate decision unit.

Evidence from the three case studies, supported this proposition. A particular example was the abolition of the TASM programme by the British government. Although an overwhelming majority of actors in the international arena, officials from the MoD and members of the opposition parties were in favour of the cancellation of the TASM as early as August 1992, the relative insulation of the ultimate decision unit, in this case the British Cabinet, from their pressure delayed the decision on the missile. Only in autumn 1993 when the abolition of the missile project had gained a majority among the actors who had direct power over the Cabinet, namely the members of the Conservative parliamentary party and key cabinet ministers, did the government abandon its plans for the development of a TASM.

The above example illustrates that the decision-making process, defined as the sequence of preference changes among the members of a network, is an important factor in our understanding of policy outcomes since not all network actors have direct influence over the ultimate decision unit. It suggests that actors who lack direct power over the ultimate decision unit can influence policy outcomes only indirectly by changing the policy preferences of those actors who have direct influence over the decision-maker. How many and which actors serve as intermediate linkages between an actor and the ultimate decision unit determines the sequence of preference changes required for him or her to exert pressure indirectly. However, if the advocates of a policy fail to change the preferences of the actors who are directly linked to the ultimate decision unit, they are not able to affect the outcome of the decision-making process.

Multilevel network theory, thus, suggests that while the structure of the network and the location of the ultimate decision unit delineates the scope of a winning coalition, it is the process which defines the preference of the winning coalition. Moreover, since the exertion of indirect pressure involves intermediaries, the number of actors advocating the eventual policy outcome typically goes beyond the winning coalition. Indeed, the findings from the three case studies suggest that the formation of a winning coalition in favour of a particular policy typically came about when a majority of national and international actors supported the policy.

Measuring Pressure in Multilevel Networks

The third theoretical development advanced in this book is a quantitative measurement of political pressure in multilevel networks in terms of the number of actors who seek to influence another actor in order to change his or her policy preferences. Multilevel network theory, thereby, goes beyond qualitative descriptive analysis. The explanatory value of this measurement is affirmed by the three case studies. In addition, the advantages of this quantitative measurement of pressure can be pointed out in comparison with the two-level game and descriptive network models in particular. Thus, unlike the notion of win-sets advanced by the two-level game, the quantitative measurement of pressure proposes a clear and consistent standard by which to assess influence. While the size of different win-sets, which is regarded as central to the explanation of outcomes in the two-level game, may alternatively be understood in terms of the number of actors who support a particular policy, the number of actors who form a coalition or the diverse political power of these actors, the degree of pressure in multilevel network theory is clearly specified.

Moreover, multilevel network theory avoids circular interpretations of influence, such as the attribution of greater size or power to the win-set whose policy is eventually adopted by the decision-makers, since it measures and hypothesizes about the influence of different coalitions independently from outcomes. In addition, the quantitative measure proposed by multilevel network theory helps to assess *changes* in the size of win-sets or coalitions and thus their ability to exert influence in the foreign policy decision-making process. In the three case studies, multilevel network theory was able to link the likelihood of changes in the actors' policy preferences to the degree of pressure to which they were exposed from other actors in the network.

The quantitative measurement of pressure in multilevel network theory also offers an alternative to the qualitative assessment of influence which has led to the proliferation of competing models and typologies in British network analysis (Jordan, 1981; Rhodes and Marsh, 1992). These typologies attempt to link policy outcomes to the preferences of specific actors or stable coalitions which are believed to dominate decision-making in distinct types of networks. One problem with this approach has been the tendency to identify new types of networks whenever empirical observations point to different dominant actors. Another is that the large number of variables or dimensions according to which network types are categorized inhibits comparisons between the ability of different actors to influence policy outcomes (Waarden, 1992). It appears that disagreement about the nature of these variables among different network models has contributed to the disillusionment with the network approach (Dowding, 1995).

Multilevel network theory avoids these problems by relating outcomes to processes rather than types of networks. Specifically, it suggests that different dominant or winning coalitions, as they have been termed in this book, can emerge within the same network as a result of the pressure which actors are able to exert on each other. This perspective not only precludes the need for new types of networks if a new winning coalition is identified, it also permits the analysis of how coalitions among actors form and transform across time and issues.

In addition, multilevel network theory offers a way of increasing the parsimony of the approach by defining power relations exclusively in terms of their direction and reducing the measurement of the influence which is exerted through these relations to a single variable, namely the number of actors who sought to change the policy preferences of a particular actor in the network at a specific point of the decision-making process. The parsimony of multilevel network theory suggests how the divisions between competing network models over the number and nature of the variables which define different types of networks and dominant coalitions may fruitfully be overcome by identifying the common ground between them. Moreover, the reduction of influence to a quantitative measurement offers an objective and consistent analysis of pressure which permits the comparison of decision-making processes in multilevel networks across countries and cases. Finally, as has been argued in the introduction of this book, the more parsimonious network approach proposed here returns to the origins of network analysis which proceeded from the assumption that the distribution of linkages among actors makes a difference in policy decision-making.

Who Defines European Foreign Policy?

Underlying the development of a multilevel network theory in this book is the question: who determines European foreign policies today? It arises from the frequently voiced assertion that multilevel decision-making processes are undermining the legitimate authority of European governments and, with it, national sovereignty. The evidence for this proposition, however, is often unsystematic or anecdotal. By integrating the national, transnational and international levels of analysis within one consistent theoretical framework, the multilevel network theory proposed and tested in the preceding chapters has sought to provide a comprehensive analysis of the ability of different actors to define European foreign policies. This section summarizes the empirical evidence presented in the three case studies in order to assess the relative influence of national and international actors in European foreign policy.

Private Actors and Pressure Groups

In the introduction of this volume, it was argued that the increasing number and diversity of private actors engaging in the making of European foreign policy has contributed to the call for further multilevel theorizing. Thus, it has been noted that international integration has led to the growing interest of national and multinational companies in the regulation of transnational contacts and exchanges. Moreover, an ever greater number of pressure groups, charities and other non-governmental organizations seek to influence foreign policies at the national and international level.

The three case studies presented in this volume have, to a certain degree, confirmed the involvement of private actors in European foreign policy. However, the empirical evidence demonstrates that the influence of private actors on the foreign policy decision-making process was as diverse as the nature of these actors. In particular, multilevel network theory shows that while some private actors, such as representatives of the European industry, had a large number of power relations and, therefore, considerable impact on the foreign policy decision-making process, others, such as the Campaign for Nuclear Disarmament, did not.

European industry representatives profited in particular from the broad scope of their relations, which included national linkages with civil servants in key ministries, party politicians and parliamentary committees as well as transnational linkages with European Union institutions. Thus, the case of European dual-use controls studies reveals how the European industry was able to form transnational coalitions and coordinate its pressure on the EU and its member states through a range of international associations, such as the Union of Industrial and Employer's Confederation of Europe and the European Defence Industry Group. The negotiations further illustrate the ability of industry representatives to serve as intermediaries between those national and international actors who did not have cross-level relations and thus were unable to draw on transnational support for their cause. German industry associations, for instance, successfully linked the growing criticism of the strict German dual-use controls among the CDU/CSU and FDP coalition parties with the international opposition to a tight European control regime. Most importantly, multilevel network theory demonstrated how these linkages undermined the traditional gatekeeper position which has frequently been attributed to national European governments.

The position of intermediary par excellence was held by the media. As transmitter of information not only among national actors, but also between national and international agents, the press and international news channels were able to influence perhaps the broadest range of actors within the multilevel European foreign policy network. However, since the media was in turn dependent upon the information and expertise from the domestic and international actors involved in each issue, it more often reflected and reinforced the prevailing foreign policy preferences within the network than a change of foreign policy.

Comparatively isolated and, therefore, weak positions among the private actors engaged in the three case studies were held by pressure groups, such as the Campaign for Nuclear Disarmament, and the German labour unions. Both sets of actors lacked a range of transnational relations which would have enabled them to form coalitions with actors who held similar policy preferences beyond their national borders or to exert pressure on those who did not. Typically, pressure groups sought to influence the foreign policy process by means of their relations with parties and parliaments. However, since an increasing number of foreign policies were ultimately decided and implemented by international organizations or in international negotiations, these pressure groups were systematically disadvantaged.

Parties and Parliaments

Crucially for an assessment of who controls European foreign policy making and whether this position is legitimate, the preceding case studies illustrate that the members of parties, parliaments and parliamentary committees, which are at the heart of the democratic process, had comparatively few power relations within the multilevel European foreign policy network. Most of their linkages were with the national executive, although it was necessary to distinguish between the government and the opposition parties in this respect. Thus, in Britain, the members of the governmental party and their MPs had a slightly greater number of power relations in the multilevel European foreign policy network due to their relations with ministers and civil servants which also functioned as intermediaries between them and other European or North American governments. Conversely, in Germany, the members of all major parties, including the SPD opposition party, had direct relations with civil servants in the key ministries because of widespread party membership among leading officials. Beyond the executive, party and parliamentary relations were limited. Specifically, these actors had few transnational linkages, the exception being with sister parties in other countries, and no relations with international organizations, such as NATO or the UN.

The preceding chapters suggest that the effect of these positions for the influence of parties and parliaments on the European foreign policy process was twofold. On the one hand, the relative isolation of party political actors in the network provided them with few relations through which they could seek to exert pressure in favour of their foreign policy preferences. Moreover, although the government parties had direct power over ministers, their influence was limited by the fact that ministers had a large number of linkages which meant that pressure from their parties represented only a small proportion of their relations within the multilevel European foreign policy network. Consequently, in the case of dual-use export regulations, cabinet ministers resisted pressure to reduce the country's dual-use export regulations, although a policy change was supported unanimously by the CDU/CSU and FDP government factions as soon as the beginning of 1993. When the ultimate decision-making authority did not rest with the parliaments, but with the cabinets or international organizations, the House of Commons or the Bundestag had very little influence over the policy process and its outcomes. On the other hand, however, the empirical findings suggest that the lack of direct linkages between party, parliamentary and committee members and transnational and international actors helped to insulate the former from the influence of the latter. Party members and parliamentarians remained primarily responsive to the preferences of the electorate, the media and their government and thus were more likely to represent the national interest, perceived as the preferences of the majority of domestic actors, than the executive.

Ministers and Ministries

Ministers and civil servants, who typically had a large number and broad range of linkages within the network, were much better placed than parliamentarians to influence the outcomes of the European foreign policy making process in the three case studies. Although the large number of their relations in the network made them accessible to pressure from a broad range of different actors, many of them not domestic, the relative impact of each was comparatively low according to the quantitative measurement for pressure proposed by multilevel network theory. The case of the British TASM project, for instance, attests that even though most of Britain's NATO partners were opposed to its nuclear policy, they were not able to increase their pressure on the British cabinet ministers above the critical 50 per cent threshold. Conversely, since the key cabinet ministers had a high number of linkages, many of them with domestic actors who continued to support the missile programme, they were able to resist the international pressure for the abolition of the missile for some time. Only after a policy change became increasingly accepted by members of the Conservative Party, in particular, in the Commons Defence Committee, the British Cabinet decided to cancel the TASM project.

Moreover, the large number of national, transnational and international relations increased the ability of governmental actors to mobilize support for their own foreign policy preferences. This capacity was enhanced by the boundary position of ministers and civil servants between the national and international arena. At the centre of crosscutting relations, both the German and British administrations were not only able to form strategic coalitions with actors who had the same foreign policy preferences, but also to exert pressure on those actors who favoured different policies. During the European negotiations over common dual-use export controls, staff from the German Economics Ministry soon joined a transgovernmental coalition among their European colleagues which favoured weak European dual-use export regulations and used their support to exert pressure on German ministers.

The boundary position of ministers and civil servants, however, could not be equated with that of a gatekeeper. As has been argued above, private companies, in particular, and their associations had a number of transnational relations which enabled them not only to exert pressure on international actors, but also to link actors across levels in transnational coalitions. Morever, the case studies also reveal that the strong position of governmental actors was frequently undermined by disagreements among different ministries or between ministers and civil servants. Since each of these actors had their own transnational and international linkages, they were often able to exert pressure for opposing policies. In the case of air strikes in Bosnia, for instance, the emerging coalition in favour of military action among the governments of the United States and Britain was repeatedly challenged by their own military officials who had formed a transnational coalition within NATO's headquarters.

International Organizations

The preceding analysis demonstrates, however, that despite internal differences and the increasing role of international organizations in the making of European foreign policy, governmental actors maintained considerable control not only over national foreign policies, but also over multilateral decisions. In particular, the empirical evidence concerning veto and blocking behaviour in the three case studies indicates that national governments retained some capacity to defer or prevent unfavourable foreign policy decisions in international organizations, such as the European Union, NATO and the UN.

Although the findings from the case studies indicate that the term veto power is misleading in that national governments were rarely able to avert a foreign policy decision when the international and transnational pressure on these organizations rose above 60 per cent of the actors linked to an organization, governments usually maintained the ability to opt-out of a multilateral decision if they did not support it. Moreover, the multinational actions investigated in the case studies did not necessarily require the participation of national governments. The common European regulations of dual-use goods export controls permitted national exceptions; air strikes in Bosnia could have been implemented by the United States and France; and NATO's abolition of its TASM requirement did not rule out the stationing of the missile in Britain.

Multilevel network theory shows, however, that the observation that governments rarely resorted to unilateral action should not lead to the conclusion that international pressures dominated in European foreign policy making. The empirical evidence generated from a multilevel network perspective rather illustrates that increasing integration allowed national actors to use their transnational linkages with international organizations to exert additional pressure on their governments when these were unresponsive. Moreover, in none of the case studies did a government decide on a foreign policy which was not also supported by a majority of domestic actors. Indeed, the three case studies presented in this book suggest that neither national, transnational nor international actors dominate the European foreign policy making process. Conversely, multilevel network theory revealed that European foreign policies typically reflected the majority of preferences among all members of the multilevel European foreign policy network.

Appendix

1. The Multilevel European Foreign Policy Network (Krahmann, 2000)

Network Outline

p.166	p.167	p.168	p.169
p.170	p.171	p.172	p.173

Codes:
*	veto positions
1	became WEU member in 1995
2	became EU member in 1995
-	self linkage

Linkages of International Organisations and Collective Decision Units:
Par:16 linkages x 651 MPs = 10416

BT: ** Dec.1990-Nov.1994: L [Par] = 24 linkages x 662 MPs = 15888

Nov. 1994-: L [Par] = 24 linkages x 672 MPs = 16128

1990-1994: 319 cdu/csu, 299 spd, 79 fdp, 8 grü, 17 pds

1994-: 295 cdu/csu, 251 spd, 47 fdp, 48 grü, 30 pds, 1 other

Weu: (PM, FS, DS, Fco, Mod in member states)+ Med + UN-SC = 47

EU-CM: (PM, FS, TS, Fco, dti in member states)+ Med + EU-Co + EP= 63

EU-Co: (PM, Cab in member states)+ Ind + Med + EU-CM + EP= 28

Nato-CM: (PM, FS, DS in member states)+ Med + Nato-Org +UN-SC= 51

Nato-Org: (Fco, Mod in member states)+ Nato-CM= 29

Osce: (PM, FS, Fco in member states)+ Med + UN-SC= 65

UN-SC: (PM, FS in member states)+ Med + UN-Org= 10

UN-Org: (Fm in member states)+ Med=22

Actor	L	PM	B-Cab	FS	DS	CE	TS	Fco	Mod	Tre	Dti	Par	Com	con	lab	lib	B-vote	B-Ind	cnd
British Prime Minister (PM)	39	-	<>	<>	<>	<>	<>	<|>	<|>	<|>	<|>	<>	<>	<>	>	>	<>		
B-Cabinet (Cab)	14	<>	-	<|>	<|>	<|>	<|>	<	<	<	<	<>	<>	<|>	>	>	<>		
Foreign Secretary (FS)	36	<>	<|>	-	<|>	<|>	<|>	<>				<>	<>	<>	>	>	<>		
Defence Secretary (DS)	29	<>	<|>	<|>	-	<|>	<|>		<>			<>	<>	<>	>	>	<>		
Chancellor of Exchequer (CE)	11	<>	<|>	<|>	<|>	-	<|>			<>		<>	<>	<>	>	>	<>		
Trade Secretary (TS)	33	<>	<|>	<|>	<|>	<|>	-				<>	<>	<>	<>	>	>	<>	<>	
Foreign Office (Fco)	30	<|>	>	<>				-	<>	<>	<>	|>	>						
Ministry of Defence (Mod)	21	<|>	>		<>			<>	-	<>	<>	|>	>					>	
Treasury (Tre)	5	<|>	>			<>		<>	<>	-	<>	|>	>						
Department of Trade (Dti)	27	<|>	>				<>	<>	<>	<>	-	|>	>					<>	
House of Commons (Par)	10416	<>	<>	<>	<>	<>	<>	<|	<|	<|	<|	-	<>	<>	<>	<>	<>	<>	<>
Committees (Com)	16	<>	<>	<>	<>	<>	<>	<	<	<	<	<>	-	<>	<>		<>	<>	
Conservatives (con)	12	<>	<|>	<>	<>	<>	<>					<>	<>	-			<>	<>	
Labour (lab)	12	<	<	<	<	<	<					<>	<>		-		<>		<>
LibDem (lib)	9	<	<	<	<	<	<					<>				-	<>		
B-Voters (B-vote)	21	<>	<>	<>	<>	<>	<>					<>	<>	<>	<>	<>	-	<>	<>
B-Industry (B-ind)	30					<>		<				<>	<>	<>	<>		<>	-	
CND (cnd)	4											<>			<>		<>		-
German Chancellor (Cha)	42	<>																	
G-Cabinet (G-Cab)	20																		
Foreign Minister (FM)	38			<>															
Defence Minister (DM)	14				<>														
Finance Minister (FM)	35					<>													
Economics Minister (EM)	13						<>												
Chancellor's Office Minister (CM)	13																		
Foreign Ministry (Fm)	33							<>											
Defence Ministry (Dm)	26								<>										
Finance Ministry (Fm)	8									<>									
Economics Ministry (Em)	31										<>								
Chancellor's Office (COf)	8																		
Bundestag (BT)	15888																		
BT-Committees (Com)	20																		
Coalition Meeting (Coa)	15																		
CDU/CSU (cdu/csu)	21																		
FDP (fdp)	19																		
SPD (spd)	20																		
Greens (gru)	17																		
PDS (pds)	17																		
Bundesrat (BR)	17																		

	Cha	G-Cab	FM	DM	FiM	EM	CM	Fm	Dm	Fim	Em	COf	BT	Com	Coa	cdu/csu	fdp	spd	gru	pds	BR	Bvg	G-Ind	Uns	G-vote
PM	<>																								
B-Cab																									
FS			<>																						
DS				<>																					
CE					<>																				
TS						<>																			
Fco								<>																	
Mod									<>																
Tre										<>															
Dti											<>														
Par																									
Com																									
con																									
lab																									
ib																									
B-vote																									
B-Ind																							<>		
cnd																									
Cha	-	<>	<|>	<|>	<|>	<|>	<|>	<|>	<|>	<|>	<|>	<>	<>	<>	<>	<>	>	>	>	>	<>				<>
G-Cab	<>	-	<>	<>	<>	<>	<>	<|	<|	<|	<|	<|	<>	<>	<	<>	<>	>	>	>	<>	<			<>
FM	<|>	<>	-	<|>	<|>	<|>	<|>	<>					<>	<>	<>	>	<>	>	>	>	<>				<>
DM	<|>	<>	<|>	-	<|>	<|>	<|>		<>				<>	<>	<>	>	<>	>	>	>	<>				<>
FiM	<|>	<>	<|>	<|>	-	<|>	<|>			<>			<>	<>	<>	>	<>	>	>	>	<>				<>
EM	<|>	<>	<|>	<|>	<|>	-	<|>				<>		<>	<>	<>	>	<>	>	>	>	<>				<>
CM	<|>	<>	<|>	<|>	<|>	<|>	-					<>	<>		<>	<>	>	>	>	>	<>				<>
Fm	<|>	|>	<>					-	<|>	<|>	<|>	<|>	>	>	>	<>	<>	>	>	>					
Dm	<|>	|>		<>				<|>	-	<|>	<|>	<|>	>	>	>	<>	<>	>	>	>			<>		
Fim	<|>	|>			<>			<|>	<|>	-	<|>	<|>	>	>	>	<>	<>	>	>	>					
Em	<|>	|>				<>		<|>	<|>	<|>	-	<|>	>	>	>	<>	<>	>	>	>			<>	<>	
COf	<>	|>					<>	<|>	<|>	<|>	<|>	-	>		>	<>	<>	>	>	>					
BT	<>	<>	<>	<>	<>	<>	<>	<	<	<	<	<	-	<>		<>	<>	<>	<>	<>	<>	<>	<>	<>	<>
Com	<>	<>	<>	<>	<>	<>		<	<	<	<		<>	-		<>	<>	<>	<>	<>			<>	<>	<>
Coa	<>	>	<>	<>	<>	<>	<>	<	<	<	<	<			-	<>	<>								<>
cdu/csu	<>	<>	<	<>	<>	<	<>	<>	<>	<>	<>	<>	<>	<>	<>	-					<>		<>	<>	<>
fdp	<	<>	<>	<	<	<>	<	<>	<>	<>	<>	<	<>	<>	<>		-				<>		<>		<>
spd	<	<	<	<	<	<	<	<	<	<	<	<	<>	<>				-			<>		<>	<>	<>
gru	<	<	<	<	<	<	<	<	<	<	<	<	<>	<>					-		<>				<>
pds	<	<	<	<	<	<	<	<	<	<	<	<	<>	<>						-	<>				<>
BR	<>	<>	<|>	<|>	<|>	<|>	<|>									<>	<>	<>	<>	<>	-		<>	<>	<>

	US-Pre	US-Nsa	US-SS	US-DS	US-TR	US-Wh	US-Sd	US-Pen	US-Cd	US-con	US-vote	Med	Weu	EU CM	EU Co	EP	Nato CM	Nato Org	Osce	UN-SC	UN Org
PM	<>											<>	<>*	<>*	>		<>*		<>*	<>*	
B-Cab												<>			>						
FS			<>									<>	<>*	<>*			<>*		<>*	<>*	
DS				<>								<>	<>*				<>*				
CE												<>									
TS					<>							<>		<>*							
Fco							<>					>	<>*	<>*				<>	<>*		<|>
Mod								<>				>	<>*					<>			
Tre												>									
Dti									<>			>		<>*							
Par												<>									
Com												<>									
con										<>		<>									
lab										<>		<>									
lib												<>									
B-vote												<>	<	<	<	<>	<			<	
B-Ind												<>			>						
cnd												<>									
Cha	<>											<>	<>*	<>*	>		<>*		<>*		
G-Cab												<>			>						
FM			<>									<>	<>*	<>*			<>*		<>*		
DM				<>								<>	<>*				<>*				
FiM												<>									
EM				<>								<>		<>*							
CM												<>									
Fm					<>							>	<>*	<>*				<>	<>*		<|>
Dm							<>					>	<>*					<>			
Fim												>									
Em									<>			>		<>*							
COf												<>									
BT												<>									
Com												<>									
Coa												<>									
cdu/csu										<>		<>									
fdp												<>									
spd										<>		<>									
gru												<>									
pds												<>									
BR												<>									

	Fr	It	Ru	Ca	Sp	Tu	Ne	Be	Sw	Au	Po	De	No	Gr	Fi	Ir	Lu	Ic
PM	◇	◇	◇	◇	◇	◇	◇	◇	◇	◇	◇	◇	◇	◇	◇	◇	◇	◇
B-Cab																		
FS	◇	◇	◇	◇	◇	◇	◇	◇	◇	◇	◇	◇	◇	◇	◇	◇	◇	◇
DS	◇	◇	◇	◇	◇	◇	◇	◇			◇	◇	◇	◇			◇	◇
CE																		
TS	◇	◇	◇	◇	◇	◇	◇	◇	◇	◇	◇	◇	◇	◇	◇	◇	◇	◇
Fco	◇	◇	◇	◇	◇	◇	◇	◇	◇	◇	◇	◇	◇	◇	◇	◇	◇	◇
Mod	◇	◇		◇		◇	◇	◇			◇	◇	◇	◇			◇	◇
Tre																		
Dti	◇	◇	◇	◇	◇	◇	◇	◇	◇	◇	◇	◇	◇	◇	◇	◇	◇	◇
Par																		
Com																		
con																		
lab																		
ib																		
B-vote																		
B-Ind																		
cnd																		
Cha	◇	◇	◇	◇	◇	◇	◇	◇	◇	◇	◇	◇	◇	◇	◇	◇	◇	◇
G-Cab																		
FM	◇	◇	◇	◇	◇	◇	◇	◇	◇	◇	◇	◇	◇	◇	◇	◇	◇	◇
DM	◇	◇	◇	◇	◇	◇	◇	◇			◇	◇	◇	◇			◇	◇
FiM																		
EM	◇	◇	◇	◇	◇	◇	◇	◇	◇	◇	◇	◇	◇	◇	◇	◇	◇	◇
CM																		
Fm	◇	◇	◇	◇	◇	◇	◇	◇	◇	◇	◇	◇	◇	◇	◇	◇	◇	◇
Dm	◇	◇		◇	◇	◇	◇	◇			◇	◇	◇	◇			◇	◇
Fim																		
Em	◇	◇	◇	◇	◇	◇	◇	◇	◇	◇	◇	◇	◇	◇	◇	◇	◇	◇
COf																		
BT																		
Com																		
Coa																		
cdu/csu																		
fdp																		
spd																		
gru																		
pds																		
BR																		

	L	PM	B-Cab	FS	DS	CE	TS	Fco	Mod	Tre	Dti	Par	Com	con	lab	lib	B-vote	B-Ind	cnd
Federal Constitutional Court (Bvg)	1																		
G-Industry (G-ind)	33																	<>	
Unions (uns)	11																		
G-Voter (G-vote)	26																		
US-President (US-Pre)	32	<>																	
US-National Security Advisor (US-Nsa)	7																		
US-Secretary of State (US-SS)	30			<>															
US-Secretary of Defence (US-SD)	24				<>														
US-Trade Representative (US-TR)	27						<>												
US-White House (US-Wh)	4																		
US-State Department (US-Sd)	26							<>											
US-Pentagon (US-Pen)	19								<>										
US-Commerce Department (US-Cd)	24										<>								
US-Congress (US-con)	15													<>	<>			<>	
US-voters (US-Vote)	X																		
Media (Med)	56	<>	<>	<>	<>	<>	<>	<	<	<	<	<>	<>	<>	<>	<>	<>	<>	<>
WEU (Weu)	47	<>*		<>*	<>*			<>*	<>*								>		
EU-Council (EU-CM)	63	<>*		<>*			<>*	<>*			<>*						>		
EU-Commission (EU-Co)	28	<	<														>	<	
EU Parliament (EP)	15																<>		
Nato-Council (Nato-CM)	51	<>*		<>*	<>*												>		
Nato-Organisation (Nato-Org)	29							<>	<>										
OSCE (Osce)	65	<>*		<>*				<>*									>		
UN-Security Council (UN-SC)	10	<>*		<>*													>		
UN-Organisation (UN-Org)	22							<|>											
France (Fr)	x	<>		<>	<>		<>	<>	<>		<>							<>	
Italy (It)	x	<>		<>	<>		<>	<>	<>		<>							<>	
Russia (Ru)	x	<>		<>	<>		<>	<>			<>							<>	
Canada (Ca)	x	<>		<>	<>		<>	<>	<>		<>							<>	
Spain (Sp)	x	<>		<>	<>		<>	<>			<>							<>	
Turkey (Tu)	x	<>		<>	<>		<>	<>	<>		<>							<>	
Netherlands (Ne)	x	<>		<>	<>		<>	<>	<>		<>							<>	
Belgium (Be)	x	<>		<>	<>		<>	<>	<>		<>							<>	
Sweden (Sw)	x	<>		<>			<>	<>			<>							<>	
Austria (Au)	x	<>		<>			<>	<>			<>							<>	
Portugal (Po)	x	<>		<>	<>		<>	<>	<>		<>							<>	
Denmark (De)	x	<>		<>	<>		<>	<>	<>		<>							<>	
Norway (No)	x	<>		<>	<>		<>	<>	<>		<>							<>	
Greece (Gr)	x	<>		<>	<>		<>	<>	<>		<>							<>	
Finland (Fi)	x	<>		<>			<>	<>			<>							<>	
Ireland (Ir)	x	<>		<>			<>	<>			<>							<>	
Luxembourg (Lu)	x	<>		<>	<>		<>	<>	<>		<>							<>	
Iceland (Ic)	x	<>		<>	<>		<>	<>	<>		<>							<>	

A/B	Cha	G-Cab	FM	DM	FM	EM	CM	Fm	Dm	Fm	Em	COf	BT	Com	Coa	cdu/csu	fdp	spd	gru	pds	BR	Bvg	G-ind	uns	G-vote
Bvg		>											<>										-		
G-ind								<>			<>	<>	<>			<>	<>	<>			<>		-	<>	<>
uns								<>			<>	<>	<>			<>		<>			<>		<>	-	<>
G-vote	<>	<>	<>	<>	<>	<>	<>					<>	<>	<>	<>	<>	<>	<>	<>	<>	<>		<>	<>	-
US-Pre	<>																								
US-Nsa																									
US-SS		<>																							
US-SD				<>																					
US-TR						<>																			
US-Wh																									
US-Sd								<>																	
US-Pen									<>																
US-Cd											<>														
US-con															<>		<>						<>		
US-vote																									
Med	<>	<>	<>	<>	<>	<>	<>	<	<	<	<	<>	<>	<>	<>	<>	<>	<>	<>	<>	<>		<>	<>	<>
Weu	<>*		<>*	<>*			<>*	<>*																	>
EU-CM	<>*		<>*			<>*		<>*			<>*														>
EU-Co	<	<													<								<		>
EP																<	<	<	<						<>
Nato-CM	<>*		<>*	<>*																					
Nato-Org								<>	<>																
Osce	<>*		<>*					<>*																	
UN-SC																									
UN-Org								<|>																	
Fr	<>		<>	<>		<>		<>	<>		<>														
It	<>		<>	<>		<>		<>	<>		<>														
Ru	<>		<>	<>		<>		<>	<>		<>														
Ca	<>		<>	<>		<>		<>	<>		<>														
Sp	<>		<>	<>		<>		<>	<>		<>														
Tu	<>		<>	<>		<>		<>	<>		<>														
Ne	<>		<>	<>		<>		<>	<>		<>														
Be	<>		<>	<>		<>		<>	<>		<>														
Sw	<>		<>	<>		<>		<>	<>		<>														
Au	<>		<>			<>		<>	<>		<>														
Po	<>		<>			<>		<>	<>		<>														
De	<>		<>	<>		<>		<>	<>		<>														
No	<>		<>	<>		<>		<>	<>		<>														
Gr	<>		<>	<>		<>		<>	<>		<>														
Fi	<>		<>	<>		<>		<>	<>		<>														
Ir	<>		<>			<>		<>	<>		<>														
Lu	<>		<>			<>		<>	<>		<>														
Ic	<>		<>	<>		<>		<>	<>		<>														

A/B	US-Pre	US-Nsa	US-SS	US-DS	US-TR	US-Wh	US-Sd	US-Pen	US-Cd	US-con	US-vote	Med	Weu	EU-CM	EU-Co	EP	Nato-CM	Nato-Org	Osce	UN-SC	UN-Org
Bvg																					
G-ind												<>			>						
uns												<>									
G-vote												<>				>					
US-Pre	-	<>	<>	<>	<>	<>		<>		<>	<	<>					<>*		<>*	<>*	
US-Nsa	<>	-	<>	<>	<>	<>				<>		<>									
US-SS	<>	<>	-	<>	<>		<>			<>		<>					<>*		<>*	<>*	
US-SD	<>	<>	<>	-	<>			<>		<>		<>					<>*				
US-TR	<>	<>	<>	<>	-				<>	<>		<>									
US-Wh	<>	<>				-				<>		>									
US-Sd		<>					-	<>	<>	>		>							<>	<>*	<|>
US-Pen	<>		<>				<>	-	<>	>		>							<>		
US-Cd				<>			<>	<>	-	>		>									
US-con	<>	<>	<>	<>	<>	<>	<	<	<	-	<	<>									
US-vote	>									>	-	<>									
Med	<>	<>	<>	<>	<>	<	<	<	<	<>	<>	-									
Weu												<>	-						<	<	
EU-CM												<>		-	<>	<>					
EU-Co												<>		<>	-	<>					
EP												<>		<>	<>	-					
Nato-CM	<>*		<>*	<>*								<>					-	<>	<	<	
Nato-Org					<>	<>						>					<>	-			
Osce	<>*		<>*		<>*							<>	>				>		-		
UN-SC	<>*		<>*									<>	>				>			-	<>
UN-Org						<|>						>								<>	-
Fr	<>		<>	<>	<>		<>	<>	<>			<>	<>*	<>*	>	>	<>*		<>*	<>*	<|>
It	<>		<>	<>	<>		<>	<>	<>			<>	<>*	<>*	>	>	<>*	<>	<>*		<|>
Ru	<>		<>	<>	<>		<>		<>			<>							<>*	<>*	<|>
Ca	<>		<>	<>	<>		<>	<>	<>			<>					<>*	<>	<>*		<|>
Sp	<>		<>	<>	<>		<>	<>	<>			<>	<>*	<>*	>	>	<>*		<>*		<|>
Tu	<>		<>	<>	<>		<>	<>	<>			<>					<>*	<>	<>*		<|>
Ne	<>		<>	<>	<>		<>	<>	<>			<>	<>*	<>*	>	>	<>*	<>	<>*		<|>
Be	<>		<>	<>	<>		<>	<>	<>			<>	<>*	<>*	>	>	<>*	<>	<>*		<|>
Sw	<>		<>		<>		<>		<>			<>		2	2	2			<>*		<|>
Au	<>		<>		<>		<>		<>			<>		2	2	2			<>*		<|>
Po	<>		<>	<>	<>		<>	<>	<>			<>	<>*	<>*	>	>	<>*	<>	<>*		<|>
De	<>		<>	<>	<>		<>	<>	<>			<>		<>*	>	>	<>*	<>	<>*		<|>
No	<>		<>	<>	<>		<>	<>	<>			<>					<>*	<>	<>*		<|>
Gr	<>		<>	<>	<>		<>	<>	<>			<>	1	<>*	>	>	<>*	<>	<>*		<|>
Fi	<>		<>		<>		<>		<>			<>		2	2	2			<>*		<|>
Ir	<>		<>		<>		<>		<>			<>		<>*	>	>			<>*		<|>
Lu	<>		<>	<>	<>		<>	<>	<>			<>	<>*	<>*	>	>	<>*	<>	<>*		<|>
Ic	<>		<>	<>	<>		<>	<>	<>			<>					<>*	<>	<>*		<|>

	Fr	It	Ru	Ca	Sp	Tu	Ne	Be	Sw	Au	Po	De	No	Gr	Fi	Ir	Lu	Ic
Bvg																		
G-ind	<>	<>	<>	<>	<>	<>	<>	<>	<>	<>	<>	<>	<>	<>	<>	<>	<>	<>
uns																		
G-vote																		
US-Pre	<>	<>	<>	<>	<>	<>	<>	<>	<>	<>	<>	<>	<>	<>	<>	<>	<>	<>
US-Nsa																		
US-SS	<>	<>	<>	<>	<>	<>	<>	<>	<>	<>	<>	<>	<>	<>	<>	<>	<>	<>
US-SD	<>	<>	<>	<>	<>	<>	<>	<>			<>	<>	<>	<>			<>	<>
US-TR	<>	<>	<>	<>	<>	<>	<>	<>	<>	<>	<>	<>	<>	<>	<>	<>	<>	<>
US-Wh																		
US-Sd	<>	<>	<>	<>	<>	<>	<>	<>	<>	<>	<>	<>	<>	<>	<>	<>	<>	<>
US-Pen	<>	<>		<>	<>	<>	<>	<>			<>	<>	<>	<>			<>	<>
US-Cd	<>	<>	<>	<>	<>	<>	<>	<>	<>	<>	<>	<>	<>	<>	<>	<>	<>	<>
US-con																		
US-vote																		
Med	<>	<>	<>	<>	<>	<>	<>	<>	<>	<>	<>	<>	<>	<>	<>	<>	<>	<>
Weu	<>*	<>*		<>*			<>*	<>*		<>*				1			<>*	
EU-CM	<>*	<>*		<>*			<>*	<>*	2	2	<>*	<>*		<>*	2	<>*	<>*	
EU-Co	<	<		<			<	<	2	2	<	<		<	2	<	<	
EP	<	<		<			<	<	2	2	<	<		<	2	<	<	
Nato-CM	<>*	<>*		<>*	<>*	<>*	<>*	<>*			<>*	<>*	<>*	<>*			<>*	<>*
Nato-Org		<>		<>			<>	<>	<>		<>	<>	<>	<>			<>	<>
Osce	<>*	<>*	<>*	<>*	<>*	<>*	<>*	<>*	<>*	<>*	<>*	<>*	<>*	<>*	<>*	<>*	<>*	<>*
UN-SC	<>*		<>*															
UN-Org	<|>	<|>	<|>	<|>	<|>	<|>	<|>	<|>	<|>	<|>	<|>	<|>	<|>	<|>	<|>	<|>	<|>	<|>
Fr	-	<>	<>	<>	<>	<>	<>	<>	<>	<>	<>	<>	<>	<>	<>	<>	<>	<>
It	<>	-	<>	<>	<>	<>	<>	<>	<>	<>	<>	<>	<>	<>	<>	<>	<>	<>
Ru	<>	<>	-	<>	<>	<>	<>	<>	<>	<>	<>	<>	<>	<>	<>	<>	<>	<>
Ca	<>	<>	<>	-	<>	<>	<>	<>	<>	<>	<>	<>	<>	<>	<>	<>	<>	<>
Sp	<>	<>	<>	<>	-	<>	<>	<>	<>	<>	<>	<>	<>	<>	<>	<>	<>	<>
Tu	<>	<>	<>	<>	<>	-	<>	<>	<>	<>	<>	<>	<>	<>	<>	<>	<>	<>
Ne	<>	<>	<>	<>	<>	<>	-	<>	<>	<>	<>	<>	<>	<>	<>	<>	<>	<>
Be	<>	<>	<>	<>	<>	<>	<>	-	<>	<>	<>	<>	<>	<>	<>	<>	<>	<>
Sw	<>	<>	<>	<>	<>	<>	<>	<>	-	<>	<>	<>	<>	<>	<>	<>	<>	<>
Au	<>	<>	<>	<>	<>	<>	<>	<>	<>	-	<>	<>	<>	<>	<>	<>	<>	<>
Po	<>	<>	<>	<>	<>	<>	<>	<>	<>	<>	-	<>	<>	<>	<>	<>	<>	<>
De	<>	<>	<>	<>	<>	<>	<>	<>	<>	<>	<>	-	<>	<>	<>	<>	<>	<>
No	<>	<>	<>	<>	<>	<>	<>	<>	<>	<>	<>	<>	-	<>	<>	<>	<>	<>
Gr	<>	<>	<>	<>	<>	<>	<>	<>	<>	<>	<>	<>	<>	-	<>	<>	<>	<>
Fi	<>	<>	<>	<>	<>	<>	<>	<>	<>	<>	<>	<>	<>	<>	-	<>	<>	<>
Ir	<>	<>	<>	<>	<>	<>	<>	<>	<>	<>	<>	<>	<>	<>	<>	-	<>	<>
Lu	<>	<>	<>	<>	<>	<>	<>	<>	<>	<>	<>	<>	<>	<>	<>	<>	-	<>
Ic	<>	<>	<>	<>	<>	<>	<>	<>	<>	<>	<>	<>	<>	<>	<>	<>	<>	-

2. Pressure and Preference Changes

European Union - The Dual-Use Control Agreement

Actor	L	T1: 14/2/92-		T2: 31/8/92-		T3: 23/10/93-		T4: 28/10/93-		T5: 12/11/93-		T6: 5/12/93-		T7: 14/4/94-		T8: 4/12/94	
		E	P	E	P	E	P	E	P	E	P	E	P	E	P	E	P
EU-Co	28	23	82%														
cdu/csu	21	1	5%	1	5%												
fdp	19	1	5%	1	5%												
Em	31	12	39%	12	39%	14	45%										
EM	35	12	34%	12	34%	13	37%	14	40%								
DM	32	10	31%	10	31%	11	34%	11	34%	12	38%						
Dm	25	10	40%	10	40%	12	48%	13	52%	13	52%	14	56%				
Coa	15	0	0%	0	0%	2	13%	3	20%	4	27%	7	47%	8	53%		
Cha	42	12	29%	12	29%	13	31%	14	33%	15	36%	18	43%	18	43%	19	45%
Cab	20	0	0%	0	0%	2	10%	3	15%	4	20%	7	35%	8	40%	9	45%
FM	38	12	32%	12	32%	13	34%	13	34%	14	37%	16	42%	16	42%	17	45%
Fm	33	11	33%	11	33%	13	39%	14	42%	14	42%	15	45%	16	48%	16	48%
BT	15888	662	4%	662	4%	10344	65%	10608	67%	10872	68%	11664	73%	11178	69%	11178	69%
Com	20	1	5%	1	5%	3	15%	4	20%	5	25%	6	30%	7	35%	7	35%
spd	20	1	5%	1	5%	1	5%	2	10%	3	15%	6	30%	7	35%	7	35%
grü	17	0	0%	0	0%	0	0%	1	6%	2	12%	5	29%	6	35%	6	35%
pds	17	0	0%	0	0%	0	0%	1	6%	2	12%	5	29%	6	35%	6	35%
BR	17	1	6%	1	6%	3	18%	3	18%	4	24%	6	35%	6	35%	6	35%
Uns	9	1	11%	1	11%	2	22%	3	33%	3	33%	3	33%	3	33%	3	33%
vote	26	2	8%	3	12%	5	19%	5	19%	6	23%	8	31%	8	31%	9	35%
Med	63	13	21%	14	22%	16	25%	17	27%	18	29%	21	33%	22	35%	23	37%
EP	15	1	7%	2	13%	2	13%	2	13%	2	13%	2	13%	2	13%	2	13%

United Kingdom - The Tactical Air-to-Surface Missile

Actor	L	T 1: 4/5/90- E	P	T 2: 11/8/91- E	P	T 3: 30/9/91- E	P	T 4: 3/10/91- E	P	T 5: 12/10/91- E	P	T 6: 8/7/92- E	P	T 7: 9/10/92- E	P	T 8: 3/7/93- E	P	T 9: 30/9/93 E	P
US-con	15	2	13%																
US-Pen	19	7	37%	7	37%														
US-Pre	32	8	25%	9	28%	10	31%												
US-SS	30	8	27%	9	30%	9	30%												
US-DS	24	8	33%	9	38%	10	42%												
US-Sd	26	8	31%	8	31%	9	35%												
Nato-CM	51	21	41%	21	41%	21	41%	21	41%										
Nato-Org	29	14	48%	14	48%	15	52%	15	52%										
Mod	21	7	33%	7	33%	8	38%	8	38%	9	43%								
DS	29	8	28%	8	28%	8	28%	8	28%	10	34%	11	38%						
Com	16	1	6%	1	6%	1	6%	1	6%	1	6%	2	13%	3	19%				
PM	39	8	21%	8	21%	8	21%	8	21%	10	26%	11	28%	12	31%	13	33%		
B-Cab	14	0	0%	0	0%	0	0%	0	0%	0	0%	1	7%	2	14%	3	21%		
FS	36	8	22%	8	22%	8	22%	8	22%	10	28%	10	28%	12	33%	13	36%		
Fco	30	8	27%	8	27%	8	27%	8	27%	10	33%	11	37%	11	37%	11	37%		
Par	10400	4814	46%	4814	46%	4814	46%	4814	46%	4814	46%	5376	52%	5736	55%	6096	59%	7536	72%
con	12	0	0%	1	8%	1	8%	1	8%	1	8%	1	8%	2	17%	3	25%	6	50%
vote	21	3	14%	3	14%	3	14%	3	14%	3	14%	3	14%	4	19%	5	24%	8	38%
Ind	30			1	3%	1	3%	1	3%	1	3%	2	7%	2	7%	3	10%	3	10%
Med	56	11	20%	12	21%	13	23%	17	30%	19	34%	20	36%	21	38%	22	39%	26	46%

Transatlantic Community - Air Strikes in Bosnia

Actor	L	T1: 15/5/92-		T2: 8/8/92-		T3: 10/8/92-		T4: 16/8/92-		T5: 29/9/92-		T6: 29/10/92-		T7: 1/12/92-	
		E	P	E	P	E	P	E	P	E	P	E	P	E	P
Med	56	11	20%												
US-con	15	3	20%												
EU-Co	28	10	36%	11	39%										
US-Pre	32	9	28%	11	34%	11	34%								
US-Wh	4	1	25%	2	50%	3	75%	4	100%						
EP	15	5	33%	6	40%	7	47%	7	47%	7	47%				
Fco	30	8	27%	8	27%	8	27%	8	27%	8	27%	8	27%		
FS	36	8	22%	9	25%	9	25%	9	25%	9	25%	9	25%	10	28%
lab	12	0	0%	2	17%	2	17%	2	17%	2	17%	2	17%	2	17%
US-DS	24	8	33%	10	42%	10	42%	11	46%	11	46%	11	46%	11	46%
EU-CM	63	25	40%	26	41%	27	43%	27	43%	27	43%	28	44%	29	46%
B-vote	21	1	5%	2	10%	3	14%	3	14%	3	14%	4	19%	4	19%
DS	29	6	21%	7	24%	7	24%	7	24%	7	24%	7	24%	7	24%
PM	39	7	18%	8	21%	8	21%	9	23%	9	23%	9	23%	10	26%
B-Cab	14	0	0%	1	7%	1	7%	1	7%	1	7%	1	7%	2	14%
Mod	21	6	29%	6	29%	6	29%	6	29%	6	29%	6	29%	7	33%
Par	10416	951	9%	1582	15%	1582	15%	1582	15%	1582	15%	1582	15%	2213	15%
Com	16	0	0%	1	6%	1	6%	1	6%	1	6%	1	6%	2	13%
con	12	0	0%	2	17%	2	17%	2	17%	2	17%	2	17%	2	17%
Weu	47	25	53%	26	55%	26	55%	26	55%	26	55%	26	55%	27	57%
Nato-CM	51	19	37%	20	39%	20	39%	21	41%	21	41%	21	41%	21	41%
Nato-Org	29	13	45%	13	45%	13	45%	13	45%	13	45%	13	45%	14	48%
Osce	65	23	35%	24	37%	24	37%	25	38%	25	38%	25	38%	26	40%
UN-SC	10	1	10%	2	20%	2	20%	3	30%	3	30%	3	30%	3	30%
UN-Org	22	8	36%	8	36%	8	36%	8	36%	8	36%	8	36%	9	41%

T8: 3/12/92-		T9: 4/12/92-		T10: 19/1/93-		T11: 23/3/93-		T12: 7/4/93-		T13: 25/4/93-		T14: 29/4/93	
E	P	E	P	E	P	E	P	E	P	E	P	E	P
3	25%												
11	46%	11	46%										
30	48%	30	48%	30	48%								
5	24%	6	29%	6	29%	7	33%						
8	28%	8	28%	9	31%	9	31%	10	34%				
11	28%	11	28%	11	28%	12	31%	13	33%	14	36%		
3	21%	3	21%	3	21%	3	21%	4	29%	5	36%	6	43%
7	33%	7	33%	7	33%	7	33%	7	33%	8	38%	9	43%
2844	21%	6096	59%	6096	59%	6096	59%	6456	62%	6816	65%	7176	69%
3	19%	4	25%	4	25%	4	25%	5	31%	6	38%	7	44%
3	25%	3	25%	3	25%	3	25%	4	33%	5	42%	6	50%
28	60%	28	60%	28	60%	28	60%	28	60%	29	62%	30	64%
22	43%	22	43%	22	43%	22	43%	22	43%	23	43%	24	47%
14	48%	14	48%	14	48%	14	48%	14	48%	14	48%	14	48%
27	42%	27	42%	26	40%	26	40%	26	40%	26	40%	27	42%
4	40%	4	40%	3	30%	3	30%	3	30%	3	30%	4	40%
9	41%	9	41%	9	41%	9	41%	9	41%	9	41%	9	41%

Bibliography

Anthony, I. (ed.) (1991), *Arms Export Regulations*, Oxford University Press, Oxford.

Atkinson, M. and Coleman, W. (1989), 'Strong States and Weak States: Sectoral Policy Networks in Advanced Capitalist Economies', *British Journal of Political Science*, Vol.19, pp.47-67.

Atkinson, M.M. and Coleman, W.D. (1992), 'Policy Networks, Policy Communities and the Problems of Governance', *Governance*, Vol.5, pp.154-180.

Baldwin, D. (1989), *Paradoxes of Power*, Blackwell, New York.

Barber, J. (1976), *Who Makes British Foreign Policy?*, Open University Press, Milton Keynes.

Barry, B. (1965/1990), *Political Argument*, Harvester Wheatsheaf, New York.

Bauer, H. and Eavis, P. (eds.) (1992), *Arms and Dual-use Exports from the EC: A Common Policy for Regulation and Control*, Saferworld, Bristol.

Benington, J. and Harvey, J. (1998), 'Transnational Local Authority Networking within the European Union: Passing Fashion or New Paradigm?', in D. Marsh (ed.), *Comparing Policy Networks*, Open University Press, Buckingham, pp.149-166.

Benson, J.K. (1982), 'A Framework for Policy Analysis', in D. Rogers et al. (eds.), *Interorganizational Co-ordination*, Iowa State University Press, Ames.

Benz, A. (1993), 'Commentary of O'Toole and Scharpf: The Network Concept as a Theoretical Approach', in F.W. Scharpf (ed.), *Games in Hierarchies and Networks*, Campus, Frankfurt/Main, pp.167-175.

Bitzinger, R.A. (1994), 'The Globalization of the Arms Industry. The Next Proliferation Challenge', *International Security*, Vol.19, pp.170-198.

Boutwell, J. (1990), *The German Nuclear Dilemma*, Brassey's, London.

Bressers, H., O'Toole, L. and Richardson, J. (1994), 'Networks as Models of Analysis: Water Policy in Comparative Perspective', *Environmental Politics*, Vol.3, pp.1-23.

Buchanan, J. and Tullock, G. (1962), *The Calculus of Consent*, University of Michigan Press, Ann Arbor.

Budd, A. (1993), *The EC and Foreign and Security Policy*, University of North London Press, London.

Cabinet Office (1992), *Questions of Procedure for Ministers*, HMSO, London.

Carver, M. (1992), *Tightrope Walking: British Defence Policy since 1945*, Random Century, London.

Connolly, W.E. (1983), *The Terms of Political Discourse*, 2nd ed., Blackwell, Oxford.

Cook, K.S., Emerson, R.M., Gillmore, M.R. and Yamagishi, T. (1983), 'The Distribution of Power in Exchange Networks: Theory and Experimental Results', *American Journal of Sociology*, Vol.89, pp.275-305.

Cooper, R.N. (1972), 'Economic Interdependence and Foreign Policy in the Seventies', *World Politics*, Vol.24, pp.161-181.

Cornish, P. (1995), *The Arms Trade and Europe*, Pinter, London.

Croft, S. and Dunn, D.H. (1990), 'The Impact of the Defence Budget on Arms Control Policy', in M. Hoffmann (ed.), *UK Arms Control Policy in the 1990s*, Manchester University Press, Manchester, pp.53-69.

Danchev, A. and Halverson, T. (eds.) (1996), *International Perspectives on the Yugoslav Crisis*, Macmillan, Basingstoke.

Daugbjerg, C. and Marsh, D. (1998), 'Explaining Policy Outcomes: Integrating the Policy Network Approach with Macro-level and Micro-level Analysis', in D. Marsh (ed.), *Comparing Policy Networks*, Open University Press, Buckingham, pp.52-71.

Defence Committee (1990), Tenth Report, *Defence Implications of Recent Events*, House of Commons Paper 320, HMSO, London.

Defence Committee (1992), First Report, *Statement on the Defence Estimates 1992*, House of Commons Paper 218, HMSO, London.

Dodd, T. (1995), *War and Peacekeeping in the Former Yugoslavia*, House of Commons Library, London.

Dowding, K. (1991), *Rational Choice and Political Power*, Edward Elgar, Aldershot.

Dowding, K. (1994), 'The Compatibility of Behaviouralism, Rational Choice and New Institutionalism', *Journal of Theoretical Politics*, Vol.6, pp.105-117.

Dowding, K. (1995), 'Model or Metaphor? A Critical Review of the Policy Network Approach', *Political Studies*, Vol.XLIII, pp.136-158.

Dowding, K. and King, D. (eds.) (1995), *Preferences, Institutions and Rational Choice*, Clarendon, Oxford.

Dunleavy, P. (1991), *Democracy, Bureaucracy and Public Choice*, Harvester, London.

Elster, J. (1986), 'Introduction', in J. Elster (ed.), *Rational Choice*, Blackwell, Oxford, pp.1-33.

Elzen, B., Enserink, B. and Smit, W. (1990), 'Weapon Innovation - Networks and Guiding Principles', *Science and Public Policy*, Vol.17, pp.171-193.

Emerson, R.M. (1962), 'Power-Dependence Relations', *American Sociological Review*, Vol.27, pp.31-40.

Evangelista, M. (1995), 'The Paradox of State Strength - Transnational Relations, Domestic Structures and Security Policy in Russia and the Soviet-Union', *International Organization*, Vol.49, pp.1-38.

Evans, P.B., Jacobson, H.K. and Putnam, R.D. (eds.) (1993), *Double-Edged Diplomacy. International Bargaining and Domestic Politics*, University of California Press, Berkeley.

Frankel, J. (1963), *The Making of Foreign Policy*, Oxford University Press, Oxford.

Freedman, L. (ed.) (1994), *Military Intervention in European Conflicts*, Blackwell, Oxford.

Gillies, D.A. (1971), 'A Falsifying Rule for Probability Statements', *British Journal for the Philosophy of Science*, Vol.22, pp.231-261.

Gow, J. (1997), *Triumph of the Lack of Will. International Diplomacy and the Yugoslav War*, Hurst&Company, London.

Green, D.P. and Shapiro, I. (1994), *Pathologies of Rational Choice Theory. A Critique of Applications in Political Science*, Yale University Press, New Haven.

Guay, T.R. (1998), *At Arm's Length: the European Union and Europe's Defence Industry*, Macmillan, Basingstoke.

Gummett, P. and Reppy, J. (1990), 'Military Industrial Networks and Technical Change in the New Strategic Environment', *Government and Opposition*, Vol.25, pp.287-303.

Halverson, T. (1996), 'American Perspectives', in A. Danchev and T. Halverson (eds.), *International Perspectives on the Yugoslav Crisis*, Macmillan, Basingstoke, pp.1-28.

Hantke, W. (1992), 'Stricter Controls on Arms Exports for Dual-use Goods: A Case Study for Drafting and Enacting Statutory Regulations' in H.G. Brauch, H. Van Der Graaf, J. Grin and W.A. Smit (eds.), *Controlling the Development and Spread of Military Technology*, VU University Press, Amsterdam, pp.257-268.

Harsanyi, J.C. (1962), 'Measurement of Social Power, Opportunity Costs and the Theory of Two-Person Bargaining Games', *Behavioral Science*, Vol.7, pp.67-80.

Heclo, H. (1978), 'Issue Networks and the Executive Establishment', in A. King (ed.), *The New American Political System*, American Enterprise Institute for Public Policy Research, Washington, D.C.

Heinz, J.P. et al. (1990), 'Inner Circles or Hollow Cores? Elite Networks in National Policy Systems', *Journal of Politics*, Vol.52, pp.356-390.

Hermann, M., Hermann, C.F. and Hagan, J.D. (1987), 'How Decision Units Shape Foreign Policy Behavior', in C.F. Hermann, J. N. Rosenau and C. Kegley (eds.), *New Directions in the Study of Foreign Policy*, Allen & Unwin, Boston, pp.309-336.

Hill, C. (1998), 'Closing the Capabilities-Expectations Gap', in J. Peterson and H. Sjursen (eds.), *A Common Foreign Policy for Europe?*, Routledge, London, pp.18-38.

Hill, C. and Wallace, W. (1996), 'Introduction. Actors and Actions', in C. Hill (ed.), *The Actors in Europe's Foreign Policy*, Routledge, London, pp.1-16.

Hollis, M. and Smith, S. (1991), 'Beware of Gurus: Structure and Action in International Relations', *Review of International Studies*, Vol.17, pp.393-410.

Jakobsen, V. (1994), *Multilateralism Matters but How? The Impact of Multilateralism on Great Power Policy Towards the Break-up of Yugoslavia*, Badia Fiesolana, San Domenico.

Jones, B. (ed.) (1994), *Politics UK*, Harvester Wheatsheaf, New York.

Jordan, G. (1981), 'Iron Triangles, Woolly Corporatism and Elastic Nets: Images of the Policy Process', *Journal of Public Policy*, Vol.1, pp.95-123.

Jordan, G. and Schubert, K. (1993), 'A Preliminary Ordering of Policy Network Labelling', *European Journal of Political Research*, Vol.21, pp.7-28.

Kassim, H. (1994), 'Policy Networks, Networks and European Union Policy-Making: A Sceptical View', *West European Politics*, Vol.17, pp.15-27.

Kavanagh, D. (1996), *British Politics. Continuities and Change*, 3rd ed., Oxford University Press, Oxford.

Kegley, C.W. Jr. and Wittkopf, E.R. (1996), *American Foreign Policy. Pattern and Process*, 5th ed., St. Martin's Press, New York.

Kenis, P. and Schneider, V. (1991), 'Policy Networks and Policy Analysis: Scrutinizing a New Analytical Toolbox', in B. Marin and R. Mayntz (eds.), *Policy Networks. Empirical Evidence and Theoretical Considerations*, Campus, Frankfurt/Main, pp.25-59.

Keohane, R.O. (1984), *After Hegemony: Cooperation and Discord in the World Political Economy*, Princeton University Press, Princeton.

Keohane, R.O. (1989), *International Institutions and State Power*, Westview Press, Boulder, Col.

Keohane, R.O. and Nye, J. (1989), *Power and Interdependence: World Politics in Transition*, 2nd ed., Scott, Foresman & Co., Glenview, Ill.

Kingdon, J.W. (1984), *Agendas, Alternatives and Public Policies*, Little, Brown and Company, Boston.

Knoke, D. (1990), *Political Networks. The Structural Perspective*, Cambridge University Press, Cambridge.

Knopf, J. (1993), 'Beyond Two-Level Games: Domestic-International Interaction in the Intermediate-Nuclear Forces Negotiations', *International Organization*, Vol.47, pp.599-628.

Krahmann, E. (2000), *Multilevel Networks in British and German Foreign Policy, 1990-95*, Ph.D. Thesis, University of London, London School of Economics, London.

Labour Party (1992), *Agenda for Change*, Labour Party, London.

Lakatos, I. (1970), 'Falsification and the Methodology of Scientific Research Programmes', in I. Lakatos and A. Musgrave (eds.), *Criticism and the Growth of Knowledge*, Cambridge University Press, Cambridge, pp.91-195.

Larkin, B.D. (1996), *Nuclear Designs: Great Britain, France and China in the Global Governance of Nuclear Arms*, Transaction Publishers, New Brunswick.

Laumann, E.O. et al. (1991), 'Organisations in Political Action: Representing Interests in National Policy Making', in B. Marin and R. Mayntz (eds.), *Policy Networks: Empirical Evidence and Theoretical Considerations*, Campus, Frankfurt/Main.

Lepick, O. (1996), 'French Perspectives', in A. Danchev and T. Halverson (eds.), *International Perspectives on the Yugoslav Crisis*, Macmillan, Basingstoke, pp.76-86.

Lijphart, A. (1971), 'Comparative Politics and the Comparative Method', *American Political Science Review*, Vol.65, pp.682-695.

Lukes, S. (1974), *Power: A Radical View*, Macmillan, London.

Lundbo, S. (1997), 'Non-Proliferation: Expansion of Export Control Mechanisms', *Aussenpolitik*, Vol.42, pp.137-147.

Mackie, T. and Marsh, D. (1995), 'The Comparative Method', in D. Marsh and G. Stoker (eds.), *Theory and Methods in Political Science*, Macmillan, Basingstoke, pp.173-188.

Macleod, A. (1997), 'French Policy toward the War in the Former Yugoslavia: A Bid for International Leadership', *International Journal*, Vol.LII, pp.243-264.

Mann, M. (1993), 'Nation-States in Europe and Other Continents: Diversifying, Developing, not Dying', *Daedalus*, Vol.122, pp.115-140.

Marin, B. and Mayntz, R. (1991), 'Introduction', in B. Marin and R. Mayntz (eds.), *Policy Networks: Empirical Evidence and Theoretical Considerations*, Campus, Frankfurt/Main, pp.11-23.

Marsh, D. (ed.) (1998), *Comparing Policy Networks*, Open University Press, Buckingham.

Marsh, D. (1998), 'The Utility and Future of Policy Network Analysis', in D. Marsh (ed.), *Comparing Policy Networks*, Open University Press, Buckingham, pp.185-197.

Marsh, D. and Rhodes, R.A.W. (1992), 'Policy Communities and Issue Networks. Beyond Typology', in D. Marsh and R.A.W. Rhodes (eds.), *Policy Networks in British Government*, Clarendon, Oxford, pp.249-268.

Mayntz, R. (1988), 'Networks, Issues and Games', in F.W. Scharpf (ed.), *Games in Hierarchies and Networks. Analytical and Empirical Approaches to the Study of Governance Institutions*, Campus, Frankfurt/Main, pp.189-209.

Mayntz, R. and Derlien, H.-U. (1989), 'Party Patronage and Politicization of the West German Administrative Elite 1970-1987 - Toward Hybridization?', *Governance*, Vol.2, pp.384-404.

McCormick, J. (1996), *The European Union. Politics and Policies*, Westview Press, Boulder, Col.

Mearsheimer, J. (1990), 'Back to the Future: Instability in Europe After the Cold War', *International Security*, Vol.15, pp.5-56.

Merritt, R.L. and Zinnes, D.A. (1989), 'Alternative Indexes of National Power', in R.J. Stoll and M.D. Ward (eds.), *Power in World Politics*, Lynne Rienner, Boulder, Col.

Moodie, M. (1995), 'The Balkan Tragedy', *Annals of the American Academy of Political and Social Science*, Vol.541, pp.101-115.

Most, B. and Starr, H. (1989), *Inquiry, Logic and International Politics*, University of South Carolina Press, Columbia.

Müller, H. and Risse-Kappen, T. (1993), 'From the Outside In and from the Inside Out. International Relations, Domestic Politics and Foreign Policy', in D. Skidmore and V.M. Hudson (eds.), *The Limits of State Autonomy. Societal Groups and Foreign Policy Formulation*, Westview Press, Boulder, Col., pp.25-48.

Nagel, J.H. (1975), *The Descriptive Analysis of Power*, Yale University Press, New Haven.

Nørgaard, A.S. (1996), 'Rediscovering Reasonable Rationality in Institutional Analysis', *European Journal of Political Research*, Vol.29, pp.31-57.

Oppenheim, F.E. (1981), *Political Concepts*, Blackwell, Oxford.

Paterson, R.H. (1997), *Britain's Strategic Nuclear Deterrent. From Before the V-Bomber to Beyond Trident*, Frank Cass & Co., London.

Peters, G. (1998), 'Policy Networks: Myth, Metaphor and Reality', in D. Marsh (ed.), *Comparing Policy Networks*, Open University Press, Buckingham, pp.21-32.

Peterson, J. (1992), 'The European Technology Community', in R.A.W. Rhodes and D. Marsh (eds.), *Policy Networks in British Government*, Clarendon, Oxford, pp.226-248.

Putnam, R. (1988), 'Diplomacy and Domestic Politics: The Logic of Two-Level Games', *International Organization*, Vol.42, pp.427-460.

Rai, M. (1993), *Britain, Maastricht and the Bomb. The Foreign and Security Policy Implications of the Treaty of the European Union*, Drava Papers, London.

Ramet, S.P. (1994), 'The Yugoslav Crisis and the West: Avoiding "Vietnam" and Blundering into "Abyssinia"', *East European Politics and Societies*, Vol.8, pp.189-219.

Ramet, S.P. (1996), *Balkan Babel. The Disintegration of Yugoslavia from the Death of Tito to Ethnic War*, Westview Press, Boulder, Col.

Rhodes, R.A.W. (1986), *The National World of Local Government*, Allen & Unwin, London.

Rhodes, R.A.W. (1992), *Beyond Westminster and Whitehall: The Sub-Central Governments of Britain*, Routledge, Allen & Unwin, London.

Rhodes, R.A.W. and Marsh, D. (1992), 'Policy Networks in British Politics. A Critique of Existing Approaches', in R.A.W. Rhodes and D. Marsh (eds.), *Policy Networks in British Government*, Clarendon, Oxford, pp.1-26.

Richardson, J.J. and Jordan, A.G. (1979), *Governing under Pressure*, Martin Robinson, Oxford.

Risse-Kappen, T. (ed.) (1995), *Bringing Transnational Relations Back In. Non-State Actors, Domestic Structures and International Institutions*, Cambridge University Press, Cambridge.

Scharpf, F.W. (ed.) (1988), *Games in Hierarchies and Networks. Analytical and Empirical Approaches to the Study of Governance Institutions*, Campus, Frankfurt/M.

Searing, D.D. (1991), 'Roles, Rules and Rationality in the New Institutionalism', *American Political Science Review*, Vol.85, pp.1249-1260.

Simpson, J. (1991), 'Nuclear Decision-Making in Britain', in H. Müller (ed.), *How Western European Nuclear Policy Is Made*, Macmillan, Basingstoke, pp.48-73.

Skidmore, D. and Hudson, V. (eds.) (1993), *The Limits of State Autonomy: Societal Groups and Foreign Policy Formulation*, Westview Press, Boulder, Col.

Skvoretz, J. and Willer, D. (1993), 'Exclusion and Power: A Test of Four Theories of Power in Exchange Networks', *American Sociological Review*, Vol.58, pp.801-818.

Smith, R. (1993), 'Resources, Commitments and the Defence Industry', in M. Clarke and P. Sabin (eds.), *British Defence Choices for the Twenty-First Century*, Brassey's, London, pp.73-89.

Statement on the Defence Estimates 1990 (1990), Cm 1022-I, HMSO, London.

Statement on the Defence Estimates 1992 (1992), Cm 1981, HMSO, London.

Statement on the Defence Estimates 1993 - 'Defending Our Future' (1993), Cm 2270, HMSO, London.

Statement on the Defence Estimates: Britain's Defence for the 90s (1991), Cm 1559-I, HMSO, London.

Vasquez, J.A. (1995), 'The Post-Positivist Debate: Reconstructing Scientific Enquiry and International Relations Theory After Enlightenment's Fall', in K. Booth and S. Smith (eds.), *International Political Theory Today*, Polity Press, Cambridge, pp.217-240.

Waarden, F. van (1992), 'Dimensions and Types of Policy Networks', *European Journal of Political Research*, Vol.21, pp.29-52.

Walker, W. and Willett, S. (1993), 'Restructuring the European Defense Industrial Base', *Defence Economics*, Vol.4, pp.141-160.

Wallace, H., Wallace, W. and Webb, C. (1977), *Policy Making in the European Communities*, Wiley, London.

Waltz, K.N. (1979), *Theory of International Politics*, Addison-Wesley, Reading, MA.

Waltz, K. (1993), 'The Emerging Structure of International Politics', *International Security*, Vol.18, pp.44-79.

Ware, R. and Watson, F.M. (1992), *The Former Yugoslavia: A Further Update*, House of Commons Library, London.

Watson, F.M. and Ware, R. (1993), *The Bosnian Conflict - A Turning Point?*, House of Commons Library, London.

Wessels, W. (1997), 'An Ever Closer Fusion? A Dynamic Macropolitical View on Integration Processes', *Journal of Common Market Studies*, Vol.35, pp.267-297.

Wheeler, N.J. (1990), 'The Dual Imperative of Britain's Nuclear Deterrent: The Soviet Threat, Alliance Politics and Arms Control', in M. Hoffman (ed.), *UK Arms Control Policy in the 1990s*, Manchester University Press, Manchester, pp.32-52.

Wildavsky, A. (1994), 'Why Self-Interest Means Less Outside of a Social Context - Cultural Contributions to a Theory of Rational Choices', *Journal of Theoretical Politics*, Vol.6, pp.131-159.

Wilks, S. and Wright, M. (eds.) (1987), *Comparative Government - Industry Relations*, Clarendon, Oxford.

Williams, P. (1990), 'British Security and Arms Control Policy: The Changing Context' in M. Hoffman (ed.), *UK Arms Control Policy in the 1990s*, Manchester University Press, Manchester, pp.11-31.

Wright, M. (1988), 'Policy Community, Policy Networks and Comparative Industrial Policies', *Political Studies*, Vol.XXXVI, pp.593-614.

Wulf, H. (1991), 'The Federal Republic of Germany', in I. Anthony (ed.), *Arms Export Regulations*, Oxford University Press, Oxford, pp.72-85.

Wulf, H. (ed.) (1993), *Arms Industry Limited*, Oxford University Press, Oxford.

Xhudo, G. (1996), *Diplomacy and Crisis Management in the Balkans. A US Foreign Policy Perspective*, St. Martin's Press, New York.

Yamagishi, T., Gillmore, M.R. and Cook, K.S. (1988), 'Network Connections and the Distribution of Power in Exchange Networks', *American Journal of Sociology*, Vol.93, pp.833-851.

Young, O. (1980), 'International Regimes: Problems of Concept Formation', *World Politics*, Vol.32, pp.331-356.

Young, O.R. (1972), 'The Perils of Odysseus: On Constructing Theories of International Relations', in R. Tanter and R. Ullman (eds.), *Theory and Policy in International Relations*, Princeton University Press, Princeton, pp.179-203.

Index

action 34
    rational 34, 38-41
    strategic 34-5
actors 29-33
    collective 30-31
    diversity 2, 4, 6-7, 155-6
    individual 30, 31
    non-unitary 31
    number 2, 4, 5-6
    private 5, 6, 17, 24, 160-61
    public 7, 17
    role 30, 32-3
    unitary 33
Aerospatiale 113-14, 134
Afghanistan 8
agency 18, 28, 36, 156
air strikes 2
Albania 6
Aldermaston Atomic Weapons
    Establishment 115, 132
Algeria 67
Amnesty International 5
Amsterdam Treaty 7
anarchy 12, 23
Angola 67
anti-proliferation agreement 46
armaments
    collaboration 51
    exports 61-3
    industry 5, 7, 8, 126
    register 53
    research and development 8
Article 223 46, 48, 50, 66
Ashdown, Paddy 85
Aspin, Les 98
Australia Group 46
Austria 7, 85, 89
authority 3, 113
    institutional 12, 26, 40
    transfer 3, 40, 50, 89

Baker, James 87, 122
Bangemann, Martin 50
Bavaria 49, 61

behaviour 21, 29, 32, 36
    types of 42
behaviouralism 13
Belgium 46, 85, 89, 92, 115-17, 125, 129-30
Benson, J.K. 18
bilateral relations 5
bipolarity 5
blocking strategy 40, 48, 98, 123, 132-3, 135, 164
Boeing 113-15, 119, 134
border controls 46
Bosnia 2, 79-112
boundary role 56, 86, 163
bounded rationality 13, 34-6
Brioni Accords 82
Britain 15, 24, 43, 46, 51-2, 54, 63, 81, 113-46
Broek, Hans van den 91
budget 114, 126
Bulgaria 6
Bundesrat 61
Bundestag 55, 72
    committees 60
    hearing 60
bureaucratism 29
Bush, George 90-91, 92, 94-5, 97, 116, 118, 122, 128

Cabinet
    British 86, 96, 99, 114, 119, 125, 127, 131, 137
    German 59, 60, 62, 63, 69
Campaign for Nuclear Disarmament 116, 121, 123-4, 160
Canada 95, 97, 117
case studies 2
    number 15
    selection 14-16
catch-all clause 52, 58, 60, 67, 70
causal hypothesis 22
causality 23
Chancellor, Germany 45, 48, 61, 118, 122
Chancellor of Exchequer 137

187

chapter outline 16
charities 6, 160
Cheney, Richard 91, 118-19, 122, 128
Chevenement, Jean-Pierre 125
Christian Democratic Union 49, 55, 58-60,
  63-4
Christian Social Union 49, 55, 58-60, 63-4
Christopher, Warren 90, 98, 100
Clinton, Bill 91, 97, 100
coalition
  change of 158-9
  formation 156, 159
  intergovernmental 86
  transgovernmental 99
  transnational 54, 66
Coalition Meeting 62
COCOM regime 47, 54, 65-6
  Interim List 58
  New Industrial List 53
collaboration 8, 19
collective decision-making unit 37, 40,
  152-4
committees 161-2
  Bundestag 60
  House of Commons 98-9, 119, 123-5,
    133-4, 135-6
common European dual-use controls 43-
  78
Common Foreign and Security Policy 7,
  57, 67
comparison 16, 159
competitiveness 47, 54-5, 60
concentration camps 90-91
Conference for Security and
  Cooperation in Europe 6-7, 82, 86-7
Congress, U.S. 90, 124, 128
Conservative Party 84, 96, 123-4, 127,
  131, 132-3, 135-7
consistency 16
constituency 39, 150
constitution 24
convention 24
Conventional Armed Forces in Europe
  Treaty 125
Cook, Karen S. 29
correlation 23
corroboration 15
cost 36-7
cost-utility 13
  calculation 34-6

optimising behaviour 35
country list 50, 64-7
criteria
  for multilevel theorizing 2
Croatia 6, 81-2
Cuba 67
culture 32
cumulation 38
Cyprus 6
Czech Republic 6-7

data
  primary 41
  secondary 41
decision-making process 13, 29, 39, 77,
  108, 143-4, 156-8
deductive analysis 22, 33
defence
  industry 63
  national 113-46
  nuclear 127-9
  policy 2
Defence Committee *see* House of
  Commons
Defence Minister, Germany 51, 57, 59, 61-
  3, 68, 118
Defence Ministry, Germany 49, 51, 56,
  61-3
Defence White Paper 127, 133, 136
delineation 21
Delors, Jacques 49, 91-2
Denmark 46, 95, 117, 125, 130
deregulation 19, 23
descriptive analysis 29, 31, 156
desire 31
differentiation
  functional 7, 18
diplomats 10
disarmament 120-21, 129-30
diversity 4
domino effect 76
dual-use controls 2, 43-78
  common European 47
  multilateral 46-7
  national 46
  tightening 45-7
dyadic power relation 22

Eagleburger, Lawrence 90, 95
Eastern Europe 44

Economics Minister, Germany 49, 51, 56,
58-9, 62, 64, 68
Economics Ministry, Germany 49, 51, 54,
56, 57-9, 62, 67
elections 21, 68, 97, 132
electorate
British 84, 98-9, 117-18, 123-4, 132
U.S. 91
embargo *see* sanctions
Emerson, Richard M. 29
enlargement 5
environment 5
essentially contested concept 21
Estonia 6
ethnic cleansing 90-91, 98
Euro-Atlantic Partnership Council 5
Euro-Corps 87
Europe
core states 3
European Armaments Agency 66
European Commission 48-51, 52, 57, 91-2
European Community/Union 1, 19, 43-78,
79, 81, 86
European Council of Ministers 50, 66, 86-
8, 92
European Defence Industry Group 66
European Economic Community 4
European Parliament 48-50, 52, 91-2
Evans, Peter 10
exogenous factors 15, 80, 114
expectations 32-3
explanation 2, 9, 156-8
export
controls 43
liberalization 53
scandals 43, 47

Federation of German Industries 47
Federation of German Wholesale and
Foreign Trade 55
Finland 7
firms 24, 30
flexible response doctrine 118
force 25
foreign affairs 3
Foreign and Commonwealth Office, U.K.
95-6, 98
Foreign Minister, Germany 57-8, 61-2, 67,
116, 118, 120, 130
Foreign Office, Germany 49, 56-7, 59, 62

foreign policy 3
Foreign Secretary, U.K. 84-5, 96, 98-9,
119
Foreign Trade Act 45, 47, 51, 64
France 43, 46, 50-51, 54, 63-4, 81, 87, 88-
92, 97, 100, 117, 121
Franco-British collaboration 115, 119, 125
Free Democratic Party 49, 55, 58-9, 63-4
functions 7, 26, 32-3
fusion thesis 1

game theory 12
gatekeeper 10, 77, 108, 144, 161, 163
generalization 31, 35
Genscher, Hans-Dietrich 116, 118, 120,
130
German Association of Machinery and
Plant Manufacturers 60
German Chambers of Commerce and
Industry 47, 60
German Labour Union Association 61
Germany 15, 43-78, 85, 89, 92, 115, 117-
18, 120-21, 125
Gillmore, Mary R. 29
Gorbachev, Mikhail 129
Green Party 64, 68
Greenpeace 5
ground troops 81, 83, 84, 94, 95, 106
Gulf War 44-5, 83

H-list 44-5, 65, 67-8
hierarchical relations 12, 18
high politics 3
horizontal relations 11, 18
hostages 81
House of Commons 84, 94, 98-9, 123,
133, 135-7
Defence Committee 119, 123-5, 134,
136
Foreign Affairs Committee 98
House of Representatives, U.S. 128
human rights 5, 81
humanitarian aid 6, 79-81, 82
Hungary 6-7
Hurd, Douglas 84-5, 96, 97-9, 119
hypotheses 2, 9, 13, 38
assessment 70-75, 101-7, 137-43, 148-
55
deterministic 15
probabilistic 15

Iceland 117, 125
IG Metall 63
independent nuclear deterrent 113
indicator 70, 101, 137
    relational 24
    relative 24-5
individualist 31
inductive analysis 22, 33
industry
    associations 19, 49, 56, 60
    British 115, 132
    European 66, 160-61
    German 43, 47-9, 55, 60
influence 21
instance 14
institutional authority 12
institutions 24, 35
intentionality 21-2, 31-4
interdependence 4, 7-8
interest 30, 33-4
    mistaken 33
    national security 44
    objective 33
    original 34
    vital 80-81
interest groups 24, 30, 160-61
intermediaries 36, 157-8, 161
Intermediate-Range Nuclear Forces Treaty
    120
internal market 43-4, 50, 54
internalization 32
international
    organizations 2, 3, 5
    regimes 5, 24, 43, 46, 65
International Red Cross 5
intervention 83
Iran 53, 65, 67
Iraq 45, 46, 47, 53, 65, 67
issue area 11, 17-19
issues 15
Italy 46, 49, 58, 63, 85, 88-9, 92, 116-17,
    125

Jacobson, Harold 10
Jordan 46
jurisdiction 3

Keohane, Robert O. 9
King, Tom 117, 125, 130, 131
Kinkel, Klaus 57-8, 61-2, 67

Knopf, Jeffrey 10
Kohl, Helmut 45, 48, 62, 118, 122
Kosovo 81

Labour Party 83, 94, 96, 98, 116, 121,
    123, 129, 132, 135
labour unions 61, 63, 68
Lance 113, 116
Länder 49, 61
last resort 122
Latvia 6
legitimate authority 3, 24, 36, 37, 87, 160
levels of analysis 2
Liberal Democratic Party 84-5, 116, 130,
    132, 135
Libya 43, 45, 47, 53, 65, 67
licensing 46, 51
linkage theory 9
Lithuania 6
Lord Carrington 80
Lord Owen 80
Lower Saxony 49, 61
lowest common denominator 43, 51

Maastricht Treaty 7, 66
Macedonia 6
Major, John 83-6, 92, 94, 96-7, 99, 126,
    133
majority 158
    qualified 40
    voting 37
Malta 6
Marsh, David 17
Martin Marietta 114-15, 134
matrix 41
Matrix Churchill affair 46
measurement
    empirical 28
    quantitative 45, 158-9
Mecklenburg-West Pomerania 49
media 60, 68, 85, 90, 99, 123, 161
Members of Parliament 84, 125
membership 30
mercenaries 6
merger 5, 8
Middle East 44, 59
military intervention 80
ministers 162-3
ministries 162-3
Ministry of Defence, U.K. 94, 96, 99, 114,

119-20, 126-7, 132-3, 136
minority 36
Missile Technology Control Regime 46, 53
Mitterrand, François 119, 121
Möllemann, Jürgen 49, 51
multilevel
    analysis 1, 8-12
    network 42
multinational corporation 4
multiplicity 4, 5-6

national
    government 113
    security 113-46
navy 126
needs 7, 25, 30, 33
negotiations 10
    Bosnian peace 81
    European dual-use 43, 47, 48, 54, 57, 63
neo-realism 23
Netherlands 46, 85, 88-9, 92, 97, 116-17, 125, 130
network
    boundaries 17, 18-19
    change 20
    definition 17
    foreign policy 19-21
    map 15, 21
    stability 18
    structure 11-12, 29, 157
    types 11, 159
network analysis
    British 2, 10-12, 17, 28, 159
    German-American 11, 12
    sociological 28
new institutionalism 13
New Strategic Concept 130
newspapers 41, 60, 85
no-fly zone 81, 94-5, 97-8
non-governmental organizations 5, 160
norms 32
North Atlantic Cooperation Council 5, 7
North Atlantic Council 86-7, 89, 94, 97, 101, 120, 122, 129
North Atlantic Treaty Organization 1, 5-6, 63, 66, 79, 86, 88-9, 99, 115, 116-17, 119-20, 129, 130-31
North Korea 53, 65, 67

Norway 117, 125
nuclear
    strategy 113, 117, 122, 130-31
    test ban 134
Nuclear Planning Group 117-18, 130
Nuclear Suppliers Group 46
Nye, Joseph S. 25

obligation 32
operationalization 41
options 35
opt-out clause 44, 58, 60, 67
organization 30
Organization for Economic Co-operation and Development 66
Organization for Security and Cooperation in Europe 6
outcome 40, 157
out-of-area 88-9

Pakistan 69
Parliaments 161-2
parsimony 16, 159
parties 30, 161-2
Partnership for Peace 5-7
Party of Democratic Socialism 64, 67-8
peace dividend 8, 115
peacekeeping 6, 8, 81
    troops 81, 100
Pentagon 91, 93, 119, 120, 128-9, 131
phases 38
pluralism 11, 29
Poland 6-7
Portugal 85-6, 89, 92
power 21-9, 36
    as currency 28
    asymmetrical 27
    bases 23
    definition 21
    degrees 27-9
    direction 26, 27-9
    dyadic 22
    hypothesis 21-3
    structure 26-7, 29
    symmetrical 27
power relation 27-8
    autonomous 26
    cumulative 26-7
    interdependent 26
    type of 22, 26-8, 36-7, 41

preference 29, 33-4, 42
 hierarchies 31
 no change 42, 71-2, 101-2, 138-9,
  150-51
 unclear or undecided 42, 72-3, 103,
  139, 152
preference changes 41-2, 73-4, 104, 139-
 40, 151
 blocked 42, 104-5, 140-43, 153-4
 proportion of 101-2, 137-8, 148-9
 sequence 80, 109, 156, 157
 series 40
 timing 142
premises 36-7
pressure 29, 36, 38, 158-9
 average 75, 106-7, 142-3, 154
 degree of 38
 groups 24, 30, 160-61
Prime Minister, U.K. 83-6, 92, 94, 96-7,
 99, 119-21, 122, 125, 133
privatization 24
production 26
proliferation 44, 46, 53
prosperity 25
public opinion 85, 90, 98-9, 118, 123-4,
 132
public-private partnership 8
Putnam, Robert 10

qualitative analysis 158
quantitative analysis 158
Quayle, Dan 122

Rabta 43, 44, 45
rational choice 12, 156
 theory 2, 12-13, 29, 35
re-nationalization 4
reductionism 27
refugees 80, 83
regimes 5, 24, 43, 46, 65
relations
 dyadic 22
 formal 11, 12
 informal 11, 12, 19
research
 period 15, 21, 34
 programme 16
resource 24
 distribution 25
 ideational 19

 intangible 25
 material 19, 23-4
 tangible 25
resource-dependence 13, 24-6, 35
reunification 49
Rexrodt, Günter 56, 58-9, 62, 64, 68
Rheinmetall 63
Rhodes, R.A.W. 17
Rifkind, Malcolm 84, 86, 96-7, 99, 114,
 131-5, 136-7
rights 32
Risse-Kappen, Thomas 9
rogue states 44, 126
roles 32-3
 boundary 39, 56
 economic 32
 multiple 33, 39
 political 32
 social 32
Romania 6
Rosenau, James N. 9
Royal Air Force 115, 125, 132, 136
Rühe, Volker 51, 57, 59, 61-2, 68
Russia 46, 82, 88, 95, 100

safe haven 2, 80, 100-101
sanctions
 economic 80, 82
 maritime 82
 military 80, 81, 98, 100
Schleswig-Holstein 61
Secretary of Defence, U.K. 84, 85, 96-7,
 99, 114, 116, 125, 130, 131-5, 136-7
Secretary of Defence, U.S. 91, 98, 118-19,
 122, 126, 128
Secretary of State, U.S. 87, 90, 94-5, 98,
 100, 122
sector 11, 17-18
 private 7
Serbia 79, 88
service 26
single market 46
six-point initiative 97
Slovakia 6
Slovenia 6, 81
Social Democratic Party of Germany 64,
 66, 68
sovereignty 8, 113, 160
Soviet Union 44, 65, 114, 118, 120-1, 125,
 129

Spain 89, 95, 100, 117
State Department, U.S. 90-92, 93, 95, 98
Stoltenberg, Gerhard 118
structure
    and agency 18, 28, 34, 157
    institutional 23, 26
    resource-dependence 25
subcontracting 19
Supergun affair 46
superpower 4
Supreme Allied Commander in Europe
    118
survival 26, 30
Sweden 7
system
    boundaries 24
    international 23
    national 23

tactical air-to-surface missile 2
Thatcher, Margaret 84, 119-21, 122, 125
theory testing 14-16, 114
Third World 44
threat 25, 114
three-level game 10
threshold 150
timing 39, 114
transatlantic community 2, 79-112
transfer of knowledge and services 52, 58,
    60, 67, 70
transgovernmental 9
transnationalism 9, 155
transsocietal 9
Treasury 127, 135
treaty 24
Treaty of Rome 46, 48, 50, 66
Trident 114, 115, 126-7, 132, 135
troops
    fighting 93
    ground 95
Turkey 6, 85, 89, 97, 117
two-level game 10, 155, 158
two-plus-four negotiations 118
typology 11, 28, 159

ultimate decision unit 36-7, 39, 48, 157,
    158
    change 36
unanimity 51
unemployment 49, 55, 61-2, 114

Union of Industrial and Employer's
    Confederation of Europe 54, 66
unions 30
United Kingdom *see* Britain
United Nations 82
    Protection Force 88, 95, 97, 101
    resolution 752 79
    resolution 770 92-4
    resolution 781 95
    resolution 836 80, 100
    Security Council 79, 86-92, 97
United States 46, 52-3, 58, 65, 79, 89, 92,
    97, 115-18, 120-22, 130-31, 134
utility 36-7

Vance, Cyrus 80
Vance-Owen Plan 97, 100-1
variable
    dependent, interdependent 14
    intermediate 157
veto 3, 37, 40, 48, 54, 82, 87-8, 91-2, 94-
    5, 100, 117, 121, 153-4, 164
Vietnam 67
voting 37

Waarden, Frans van 18
Warsaw Treaty Organization 5, 44, 53
WE-177 free-fall nuclear bomb 115, 123,
    127, 129-30
weapons
    biological 53
    chemical 43, 45, 53
    export controls 61-2
    mass destruction 66
    nuclear 53
Weapons of War Act 45, 62
welfare 25, 30, 33
West European Armaments Group 65
Western European Union 6, 65, 86-7, 89
winning coalition 39-40, 99, 109, 156-9
win-set 10, 158
working group
    intergovernmental 50
    national 62
Wörner, Manfred 117

Yamagishi, Toshio 29
Yugoslavia 6, 44, 82

Zangger Committee 46

*For Product Safety Concerns and Information please contact
our EU representative GPSR@taylorandfrancis.com Taylor & Francis
Verlag GmbH, Kaufingerstraße 24, 80331 München, Germany*

T - #0117 - 160425 - C0 - 219/152/11 - PB - 9781138716063 - Gloss Lamination